Klaus Kircher

Chemical Reactions
in Plastics Processing

Klaus Kircher

Chemical Reactions in Plastics Processing

With 44 Figures and 51 Tables

Hanser Publishers, Munich Vienna New York

Distributed in the United States of America by
Macmillan Publishing Company, New York
and in Canada by
Collier Macmillan Canada, Inc., Ontario

English edition of
"Chemische Reaktionen bei der Kunststoffverarbeitung"
by Dr. Klaus Kircher

Translated by Dr. Carla B. Grot, Chadds Ford, PA 19317, U.S.A.

Distributed in USA by
Scientific and Technical Books
Macmillan Publishing Company
866 Third Avenue, New York, N.Y. 10022

Distributed in Canada by
Collier Macmillan Canada, Inc.
1200 Eglington Ave. E, Suite 200, Don Mills, Ontario M3C 3N1 Canada

Distributed in all other countries by
Carl Hanser Verlag
Kolbergerstrasse 22
D-8000 München 80

The use of general descriptive names, trademarks, etc., in this publication, even if the former are not especially identified, is not to be taken as a sign that such names, as understood by the Trade Marks and Merchandise Marks Act, may accordingly be used freely by anyone.

While the advice and information in this book are believed to be true and accurate at the date of going to press, neither the authors nor the editors nor the publisher can accept any legal responsibility for any errors or omissions that may be made. The publisher makes no warranty, express or implied, with respect to the material contained herein.

CIP-Kurztitelaufnahme der Deutschen Bibliothek

Kircher, Klaus:
Chemical reactions in plastics processing / Klaus
Kircher. [Transl. by Carla B. Grot]. – Munich ;
Vienna ; New York : Hanser ; New York : Macmillan, 1987.
    Dt. Ausg. u. d. T.: Kircher, Klaus: Chemische
Reaktionen bei der Kunststoffverarbeitung
    ISBN 3-446-14279-7 (Hanser)
    ISBN 0-02-947520-1 (Macmillan)

ISBN 0-02-947520-1 Macmillan Publishing Company, New York
Library of Congress Catalog Card Number 86-602089

Printed in the Federal Republic of Germany

Preface

A polymer of high molecular weight is the most commonly used raw material for manufacturing molded articles and semifinished goods. The polymer is heated, molded while still in the melt, and subsequently cooled. This molding process is strictly a physical process; chemical reactions are not considered.

Approximately 10% (by weight) of the molded articles and semifinished goods produced in the polymer processing plant undergo a chemical synthesis during processing; the processor does not start with a finished polymer of high molecular weight but manufactures the polymer during the molding process. Thus, the polymer processor performs the task of a chemist. Other methods which employ polymers in the manufacture of molded articles and semifinished goods start with high molecular weight intermediates, but the molecular weight is increased even further during the molding process, and this alters the properties of the original material. Cross-linking is one of these methods. An additional 14% of all molded articles and semifinished goods are produced by these methods. Altogether, one-fourth of all processing methods employ a chemical reaction. Even during the molding of thermoplastics it is possible for a chemical reaction to occur, for instance, an undesirable decomposition which may lead to end products of diminished quality. The processor must be aware of these undesired chemical reactions.

This book discusses every polymer processing method in which a chemical reaction is an important part of the process. I have relied to a great extent on the following books and publications and have quoted freely from the original texts.

- *Menges, G.*   Werkstoffkunde der Kunststoffe, Carl Hanser Verlag, München, 1979
- *Vieweg, R., u. a.*   Kunststoff-Handbuch (11 Bände), Carl Hanser Verlag, München, 1963 bis 1975

Bd. I: Grundlagen, Aufbau, Verarbeitung und Prüfung der Kunststoffe

Bd. II: Polyvinylchlorid

Bd. III: Abgewandelte Naturstoffe

Bd. IV: Polyolefine

Bd. V: Polystyrol

Bd. VI: Polyamide

Bd. VII: Polyurethane
(also available in English version: *Oertel*, Polyurethane Handbook)

Bd. VIII: Polyester, gesättigte Polyester, ungesättigte Polyester, Polycarbonate

Bd. IX: Polymethacrylate

Bd. X: Duroplaste

Bd. XI: Polyacetale, Epoxidharze, fluorhaltige Polymerisate, Silikone usw.

- *Houwink, R.,*   "Chemie und Technologie der Kunststoffe", 4. Aufl., Akad. Verlags-
  A. J. Stavermann   gesellschaft, Leipzig, 1962
- *Saunders, J. H.,*   "Polyurethanes, Chemistry and Technology". Interscience Publishers,
  K. C. Frisch   John Wiley & Sons, New York – London, Bd. 1, 1962

- *Boenig, H. V.* "Unsaturated Polyesters: Structure and Properties". Elsevier Publishing Comp., Amsterdam – London – New York, 1964

- *Elias, H. G.* "Neue polymere Werkstoffe", Carl Hanser Verlag, München – Wien, 1975

- *Jahn, H.* "Epoxidharze", VEB Deutscher Verlag für Grundindustrie, Leipzig, 1969

- *Voigt, J.* "Die Stabilisierung der Kunststoffe gegen Licht und Wärme", Springer Verlag, Berlin – Heidelberg – New York, 1966

- *Hofmann, W.* "Vulcanization and Vulcanizing Agents", MacLaren and Sons, Ltd., London and: Palmeton Publishing Co., New York

This book has been written during my employment with the Institut für Kunststoffverarbeitung (IKV) in Industrie und Handwerk an der Rheinisch-Westfälischen Technischen Hochschule, Aachen.

It describes parts of IKV's newest research results in the area of UP resin processing, of UP resin processing, synthesis of PUR foam articles, and cross-linking of polyethylene. I am deeply grateful to my many colleagues at IKV; Dr. E. Roth, Dr. B. Franzkoch, Dr. H. Schwesig, and Dr. A. Behmer. Part of this research has been funded by Arbeitsgemeinschaft Industrieller Forschungsvereinigungen (AIF) and Deutschen Forschungsgemeinschaft (DFG).

Finally I would like to express my sincere gratitude to Professor Dr. Ing. G. Menges, director at IKV, for encouraging me to start this book and his advice and assistance which allowed me to write it.

Spring 1987   Dr. Klaus Kircher

Contents

1 Chemical Reactions in the Polymer Processing Plant

In manufacturing parts and semifinished shapes from synthetic polymers, one usually melts a powdery or granular raw material; the melt is then shaped and subsequently cooled. During this process the synthetic polymer supplied by the chemical manufacturer is transformed into a different physical state; it is then returned to its original physical state by cooling. In the ideal case, such a shaping process is strictly a physical process which is not accompanied by a chemical reaction. But if one looks at all possible processes for the manufacturing of parts and semifinished shapes from synthetic polymers, including both synthetic macromolecular polymers and elastomers, one will find quite a few processes in which the polymer processor determines the chemical structure of the end product simultaneously with the molding process. The products which were used by the polymer processor and the end product differ in their chemical structure. The product which is supplied by the chemical manufacturer undergoes a chemical reaction in connection with the molding process. The manufacturing processes, which are characterized not only by physical change but also by targeted chemical reactions, can be divided into two groups:

(a) Manufacturing processes which, starting with a low molecular weight monomer, synthesize a polymer during the molding process.

(b) Manufacturing processes in which a high molecular weight compound is transformed into a new compound by molecular enlargement through chemical reactions. The high molecular weight compounds used in these processes may themselves be useful plastic materials, for instance, polyethylene. Through chemical reactions a new polymer with certain targeted properties is being produced.

Polymer processing includes, therefore, part of the polymer synthesis. The distinction between polymer producers and polymer processors is not clear-cut; the transition is rather vague. In many cases the processor completes the polymer synthesis which was started by the raw material producer. He therefore takes over tasks which according to popular belief belong to the chemist. Chemical reactions in the polymer processing plant are not limited to synthesis alone. Other chemical reactions also play a more or less important role. These chemical reactions sometimes occur during the manufacturing of parts and semifinished shapes; sometimes they are part of other production steps.

A chemical reaction which frequently occurs (although to a small extent) is molecular degradation. In a few cases this molecular degradation is desired; in most cases it is an unwanted side reaction which sometimes presents processing problems and at other times has detrimental effects on the properties of the end product. The processor has to minimize or eliminate these chemical reactions by optimizing the processing parameters. Molecular degradation is possible not only during the shaping process but also during storage and use of the parts. Even if these reactions do not occur in the processing plant, the polymer processor is still sometimes able, in a few cases (for instance, in the case of polyvinyl chloride) to influence the extent and outcome of subsequent reactions through the use of a suitable additive.

Chemical reactions also occur during surface treatment and during the joining of parts and semifinished shapes. Two-component adhesives, for instance, represent reactive components or component mixtures which will become a high molecular weight compound during the joining process. Sometimes, two-component adhesives are used which are very similar in their principal chemical structure to the reactive resins which are used in the manufacturing of molded parts.

Sometimes surface enhancement of the plastic parts includes several chemical reactions. Frequently this step is preceded by preparation of the surface. For instance, oxidation of the polymer during this treatment will produce polar structures and partial oxidation will roughen the surface. For surface coating one uses lacquers, which are partially reactive systems. Sometimes these reactive lacquer systems show similarities to thermosets which are used for manufacturing of the parts.

Table 1 Chemical reactions in the realm of the polymer processor

Intended chemical reactions during the processing period	– Synthesis of the polymer out of low molecular weight components – Polymer synthesis by modification of high molecular weight components – Degradation of exceedingly high molecular weight polymers (for ex.: mastication of natural rubber) – Surface modification and coating – Joining by reactive adhesives
Unwanted chemical reactions in the form of side reactions during the processing period	– Decomposition, depolymerization of the molecule – Oxidation
Chemical reactions which occur outside of the processing plant but which are still controllable by the processor through the use of additives	– Decomposition (caused by exposure to energy; remedy: addition of stabilizers) – Oxidation on exposure to weathering or fire (remedy: employment of oxidation inhibitors and fire retardant substances)

Polymer processing is a technical process during which very dissimilar chemical reactions can occur. The present text is limited to those reactions which occur during the manufacturing process of parts and semifinished shapes; chemical reactions which occur during the joining and enhancing process will not be discussed.

2 Polymers Produced by the Plastics Processor – An Overview

Table 2 contains a summary of synthetic polymers which are produced in the processing plant during the manufacturing of parts and semifinished shapes by synthesis of macromolecular products. The variety of polymers produced by the processor include many dissimilar products which cannot be identified with a special group of substances. There are several ways of classifying these polymers:

- Economic importance of the polymer
- Physical properties (thermoplastics – thermosets)
- Molecular weights of the original components
- Type of chemical reaction used in the synthesis

Synthetic polymers which are produced through chemical reactions include polymers that differ greatly in their importance. For instance, the cross-linking of elastomers of all types in terms of volume is a very important reaction. The synthesis of polyurethanes (PUR) one of the most widely produced types of polymer and one which is almost exclusively synthesized by the polymer processor) is also a very important chemical reaction in the plastics processing plant. To a large extent, therefore, the large number of plastics processors and not the chemical companies are the producers of polyurethanes.

Table 2 Polymers produced in the polymer processing plant

Synthesis of the polymer from low molecular weight components	Synthesis of the polymer from high molecular weight intermediates by chemical modification (increase in the size of molecule)
non cross-linked polymers: - Polyamide-6 - PMMA - Polycarbodiimide **cross-linked polymers:** - UP resins (cured) - Polydiallyl esters - specialty plastics of high heat distortion temperature - Phenoplasts - Aminoplasts - PUR - Polyisocyanurate - Epoxy resins (cured)	- Cross-linked thermoplasts (especially PE, PP, ethylene-vinyl acetate copolymers) - Elastomers (cross-linking of (ethylene-like) unsaturated natural rubbers, polyethylene sulfochloride, polychloroprene, polyurethane-elastomers, fluoroelastomers, polyisobutylene)

Some polymers which are produced in very small quantities are also made by molding processes which are accompanied by chemical reactions. For instance, polymers with high heat distortion temperatures are manufactured in such small amounts that they are omitted from production or manufacturing output tables. Similarly the polydiallyl esters are specialties which are theoretically very interesting but whose production output is of little importance. The plastics processor produces uncross-linked thermoplastic polymers as well as cross-linked elastic, or thermosetting molding materials.

Polyamide-6 and PMMA are thermoplastic materials which are produced by the manufacturer through reaction casting only in connection with the manufacturing of parts or semifinished shapes. Both polymers are also processed by injection molding and extrusion and the processor obtains the finished materials from the producer of raw materials.

Molded parts and semifinished goods made from polyamide-6 and PMMA can, therefore, be manufactured by means of two fundamentally different processes, each of which offers certain advantages. Thermoplastic processing of a preprocessed polymer can present difficulties for the production of very bulky objects made from these materials. On the other hand, reaction casting can be used to manufacture objects of any size. Parts made from PMMA are usually of higher quality (if the casting process is used), especially in respect to surface finish. It is a disadvantage that the reaction casting processes are difficult to run automatically and are therefore more labor intensive. Reaction casting processes have a distinct advantage over the process that uses finished polymers; they are controllable. In each step the polymer processor theoretically is able to change the composition of the mixture and consequently, up to a certain point, vary the properties of the polymer. This allows the processor to be independent of the variety of polymers that are offered by the chemical companies. It is believed that the processor also produces other thermoplastic polymers by reaction casting. Only a few chemical syntheses are performed in the processing plant, because of the difficulty in handling many of the polymer precursors and additives which are necessary for synthesis.

Processes which yield crosslinked polymers are being used much more than reaction casting processes for thermoplastic polymers. Reaction casting processes which manufacture thermoplastic polymers are in competition with other manufacturing processes; but parts and semifinished goods from cross-linked polymers can be manufactured only by reaction casting. Cross-linked polymers cannot be shaped plastically, and therefore the plastic shaping has to be done in a chemical state which allows molding; this means it is done in the prepolymer state or before the polymer is cross-linked.

Manufacturing parts from cross-linked polymers requires a chemical reaction. The cross-linked polymer can be either formed from low molecular weight reagents or produced from a plastic or thermoplastic polymer by starting new chemical bonds between various chain molecules. Sometimes, very low molecular weight reagents are used for the production of cross-linked polymers in the mold.

In the production of epoxy resins, for example, by hardening of cycloaliphatic resin, components with very low molecular weights are used. Similarly, in the hardening of unsaturated polyester (UP) resins, diallyl phthalate resins and pheno- and aminoplasts use resin components which have low or medium molecular weights. Only through a chemical reaction will they form a polymer; usually the molecular weights of the polyester components in UP resins are not higher than 4000. The same is true for the polyols in the resin which is used for the production of polyurethane.

In addition to cross-linked polymers which have been produced from low to medium molecular weight compounds, other polymers are available which are produced by cross-linking high molecular weight compounds. In a few cases these high molecular weight components are already technically useful polymers which are only modified by the cross-linking, for example, polyethylene. In other cases (like rubber), the high molecular weight starting material is not useful as a polymer for all practical purposes. Only the cross-linking process converts it into a polymer (e.g. rubber). The cross-linking reaction of only a few chain segments of the high molecular weight compound alters the properties of the rubber substantially.

Synthesis of polymers in the polymer processing plant is not limited to a specific variation of a chemical process. All three major methods of synthesis of macromolecules (polymerization, polycondensation, and polyaddition) are also used in the processing plant. In this text the different processing procedures are organized according to the chemical reactions which occur during synthesis. This makes it possible to summarize common characteristics.

3 Importance of the Polymer Synthesis in the Polymer Processing Plant

Up to 1920 practically every type of polymer processing involved a chemical reaction. Only with the development of the thermoplasts and the increasing perfection of the processing plant did the commercial importance of the reaction resins diminish. Today, a large portion of the polycondensates, so dominant in the earlier years, are used in processing applications which have little in common with polymer processing (treating of textiles, paper industry, paint and varnish industry, etc.); on the other hand, even today a considerable amount of the polymerizate is used by the fiber industry. Therefore, the data which are published in regard to production and consumption often do not report the true values for the polymer quantities processed in the processing plant. It is worth mentioning that polymer processing employs only a small portion of the phenoplasts, aminoplasts and epoxy resins; the bulk of it is used as laquers, glues and varnish and sizing material or as encapsulating compounds.

Table 3 lists the most important polymers whose final composition is only first synthesized by the processor; the same table shows the portion used by the processor in comparison to the total production. In contrast to polyurethane which is synthesized almost exclusively by the polymer processor, only one tenth of the urea and melamine resins are fabricated to cured objects.

Polymers modified by a chemical reaction, particularly cross-linked, have to be added to the list of plastics (which are) synthesized by the processor. Approximately 2 million tons of polymers are synthesized annually in polymer processing plants in Western Europe; but in addition to that approx. 3 millions of tons of elastomers (mainly in synthetic rubber factories) are produced by cross-linking of natural rubber. Therefore, the multitude of polymer processing plants represents a significant independant chemical industry.

Table 3 Polymers manufactured in the polymer processing plant by chemical reaction (not counting any specialty products) and their percentage of the total production respectively

Polymer	Percentage (by weight) of the total production carried out by the polymer processor.
Polyamide	trivial amounts
UP-resin, cured, without reinforcement	80
Polyurethane	95
Epoxy resins, cured	20
Phenolresins, cured	20
Urea and melamine resins, cured	10
Polyacrylate and methacrylate	40

Table 4 Organic material manufactured as modified by means of a reaction as percentage (by weight) of the total volume of organic polymer material

Type of material	Percentage in % by weight of the total volume
Melt fabricable plastic	75
Reaction molded plastic	10 ⎫
Cross-linked plastic as rubber	15 ⎭ 25

Table 4 shows the (percentage) break-down of the total amount of organic polymers: In approx. one fourth of the plastics (including elastomers) the characteristic properties are caused by chemical reactions designed by the processor. The processor is able to intervene in any phase of the reaction, be it by varying the ratio of the components or by altering the conditions of the reactions.

Bibliography to Chapter 3

[1] *N. N.:* Kunststoffe 71 (1981), S. 193.
[2] *N. N.:* Mod. Plast. Intern. 11 (1981), S. 33.
[3] *N. N.:* Der Kunststoffmarkt in Westeuropa, Kunststoffe 69 (1979), S. 496.
[4] *Günther, W.:* Der Kautschukmarkt in Westeuropa, Kunststoffe 69 (1979), S. 535.

4 General Methods of Polymer Synthesis and their Applicability in the Processing Plant

Macromolecular polymers are either fully synthesized from low molecular weight reagents or semisynthesized by using suitable natural substances. Polymers made from natural substances include the various cellulose esters which are manufactured solely by the chemical industry.

On the other hand, polyisoprene in the form of rubber is cross-linked solely by the polymer processor. Casein is another natural raw material which can be used for polymer synthesis, but today this process is no longer important.

Synthetic macromolecular molding compounds are synthesized according to three different methods:

- Polymerization
- Polycondensation
- Polyaddition

Each of these methods can be used to manufacture parts and semifinished goods in the polymer processing plant.

It is possible to manufacture PMMA and polyamide-6 into molding compounds by polymerization. Curing of unsaturated polyester resins and diallyl-ester resins represents polymerization reactions. Curing of pheno and amino resins as well as carbodiimide formation represents polycondensation reactions. Curing of epoxy-urethanes and synthesis of polyurethanes and polyisocyanurates represent polyaddition.

These three different methods are distinguished by specific characteristics. Polymerizations are chain reactions free of cleavage products; the mechanical properties of the product are influenced by the kinetic chain length. Polycondensation is characterized by producing a cleavage product of low molecular weight which is eliminated as a gas. Therefore, an extensive preliminary reaction between the components is necessary in order for only a few molar reactions to take place during curing and in so doing minimize voids in the molded part caused by gaseous products. Elimination of gaseous compounds does not present any problems for methods which produce foams; the cleavage product can be used as an auxiliary (expanding) agent.

In the chemical plant, industrial synthesis by polymerization of molding compounds of high molecular weight uses different types of initiation:

- Radical polymerization
- Anionic polymerization
- Cationic polymerization
- Coordinative, metalloorganic polymerization

Usually, ionic and coordinative initiated polymerizations are very sensitive toward impurities of the reactants, and the necessary initiators are difficult to handle; this limits their use in the processing plant. Radical polymerizations proceed relatively "safely"; therefore, in the processing plant most of the polymerizations are initiated by free radicals. Only one ionic polymerization belongs to the variety of polymer processing; the anionic polymerization of ε-caprolactam to polyamide.

Table 5   Classification of polymers

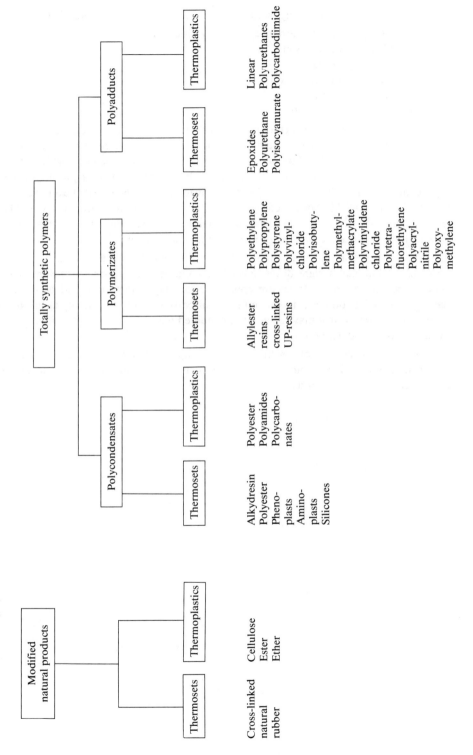

Polymerization methods may be classified further according to technical criteria:
- Bulk polymerization (substance polymerization)
- Gas phase polymerization
- Solution polymerization
- Precipitation polymerization
- Emulsion polymerization
- Suspension polymerization

Manufacture of parts and semifinished goods, in conjunction with polymer synthesis, is limited to monomers which are liquid or solid at room temperature and can be polymerized by mass polymerization. Solution, emulsion, and suspension methods, which are very important in the chemical industry, cannot be used here, since the auxiliary agent is present in large quantities (solution, emulsion, and suspension liquid) and cannot be allowed to remain in the product. Methods used by the polymer processor allow only those auxiliary agents to be used which, after the part is formed, are able to remain in the polymer without showing long-lasting effects on the properties of the molded compound.

Polycondensation and polyaddition do not differentiate between basically different mechanisms of initiation; here, different catalysts come into play. The same reasons that are valid for polymerization methods are applicable for polycondensation and polyaddition reactions; in conjunction with the manufacture of parts, only those auxiliary agents are allowed to be present which may remain in the molded part. This makes "reaction in mass" the only possible industrial process.

Liberation of the heat of reaction and reaction-caused volume contraction are common to all methods of the molding process that include a chemical reaction. These effects are present in all three methods.

5 Heat Release and Volume Contraction during the Synthesis of Polymers

One can describe most chemical reactions as processes wherein the internal energy and the specific weight of end products differ from those of the original materials. Polymer syntheses are mostly exothermic and are accompanied by a volume contraction of the reactants. The heat of reaction and the reaction related volume contractions determine, to a great extent, the feasibility of the chemical reaction and its limitations in the processing plant.

Table 6 shows typical values for volume contraction and heat of reaction. The values vary substantially. For epoxy resins, no value is given for heat release and only 0.5% for volume contractions, on the other hand, polymerization of methyl methacrylate shows a 23% volume contraction and 54 kJ/mole heat release. Polymerization of styrene shows an even greater heat release (relative to weight) but only 17% volume contraction.

Table 6 Contraction and release of heat during polymerization and polycondensation [1]

Monomers	Contraction (%)	Heat of Polymerization (kJ/mol)
Phenol Formaldehyde	0,5	
Epoxy resins	0,5	
Butylmethacrylate	15	57
Styrene	17	71
Ethylmethacrylate	18	59
Methylmethacrylate	23	54
Vinylacetate	22	88
Acrylonitrile	26	77

Volume contraction and heat of reaction do not present large problems during the curing of epoxy resins; however, in a multitude of cases, special precautions are required during polymerization of styrene or methyl methacrylate.

Highly exothermic reactions lead to high mass temperatures; the reaction then stops or decomposition occurs. At high temperatures polystyrene and polymethyl methacrylate decompose to monomeric components, while at low temperature the polymers are being formed from monomers. If one plots the equilibrium vapor pressures of the monomer (a) over the polymer and (b) over the monomer as a function of the temperature one will see two divergent curves (Figure 1). The intersection of the two curves indicates the temperature at which decomposition reactions occur at the same rate as the monomer consuming reaction. According to Figure 1 for PMMA, this temperature is comparatively low at about 200 °C (according to reference [2] this temperature is 220 °C).

If polymerization of methyl methacrylate takes place under adiabatic conditions and at an initial temperature of 50 °C the temperature of the polymer quickly rises to a temperature level at which rapid decomposition takes place.

Since polymers have a low thermal conductivity, a certain amount of heat of polymerization will always stay inside the polymer. During the manufacture of thick-walled parts, an almost adiabatic reaction occurs at points far away from the wall, especially if the rate of reaction is very high. Hot spots and large temperature differences are undesirable for sever-

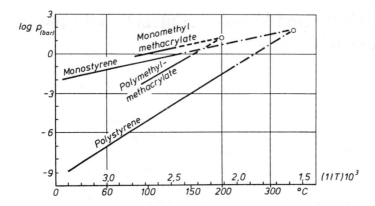

Figure 1   Thermodynamic stability of polystyrene and polymethyl methacrylate. Vapor pressures of the monomers over the polymerizing monomers and over the polymers. (3)

al reasons and have to be avoided as much as possible since they can adversely influence the quality of the plastic parts.

The heat released is directly proportional to the molar amounts of reacted groups. The heat of crystallization has to be included when a liquid monomer produces a crystalline polymer. A reduction in the heat of reaction is possible only if the molar amount of reactive groups is decreased. During polymerization of methyl methacrylate, partial heat removal in a separate prepolymerization step is possible. This preliminary reaction is possible only to a certain extent since the increasing polymer content quickly limits the casting capability of the polymerization mixture. During polycondensation and polyaddition, at first only low molecular weight intermediates are formed, and they can be processed easily even after a high degree of conversion. It is possible in these cases to remove a substantial amount of the heat of reaction through preliminary reaction, so that fewer problems are encountered during the curing process.

Exceptions are reaction molding of polyamide and polycarbodiimide. Both of these processes make a prepolymer unnecessary since the heat of reaction is very small.

The use of fillers to reduce the heat released per mass unit proves to be very successful. The curing of UP resin normally uses a large amount of fillers (partly in the form of reinforcing fibers) which automatically lower the heat of reaction during the process. The heat of reaction presents a serious problem during the reaction casting of PMMA; mineral fillers are seldom used here, and therefore a long polymerization time is required.

Like the heat of reaction, reaction-caused volume contraction is proportional to the molar conversion of the components. Therefore, a preliminary reaction will reduce a proportional amount of volume contraction. Molding materials with a limited molar conversion during curing will exhibit only a small amount of shrinkage. Methyl methacrylate and unsaturated polyesters (UP) resins exhibit a large amount of shrinkage and preliminary reactions are possible only to a very limited extent.

A reduction of shrinkage, without altering the chemical composition of the material, is possible only through the use of fillers. The heat of polymerization can be removed if the reaction occurs at a very slow rate. Shrinkage, on the other hand, is independent of the rate of reaction.

During curing of unsaturated polyesters (UP) resins, manipulation of the volume contraction of the molded part is possible through formation of evenly distributed microcavities; this will result in a molded part which does not show any shrinkage in its outer dimensions (low-profile resins).

Bibliography to Chapter 5

[1] *Kautter, C. Th.:* Kunststoff-Handbuch Bd. IX, »Polymethacrylate«. Herausg. R. Vieweg und E. Essen, Carl Hanser Verlag, München, 1975, S. 17.

[2] *Lenz, R. W.:* »Organic Chemistry of Synthetic High Polymers« – Interscience Publishers. New York, London, 1967, S. 742.

[3] a) *Houwink, R., A. J. Stavermann:* »Chemie und Technologie der Kunststoffe«, 4. Auflage. Akademische Verlagsgesellschaft, Leipzig, 1962, S. 175.

    b) *Boundy, R. H., R. F. Boyer:* »Styrene, its Polymers, Copolymers and Derivatives«. Reinhold Publishing Corp., New York, 1952.

    c) *Ivim, K. J.:* Trans. Faraday Soc. 51 (1955), S. 1273.

6 Processing in Conjunction with a Polymerization Reaction

6.1 General Concepts of Polymerization Reactions

6.1.1 Course of Free Radical Polymerization

A free radical polymerization requires a polymerizable compound and a source of free radicals which can supply initial radicals.

In principle, all compounds which have a certain type of olefinic double bond are polymerizable:

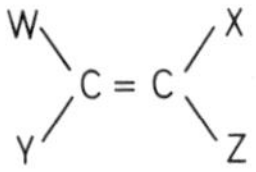

$$\begin{array}{c}W\\ \diagdown\\ \diagup\\ Y\end{array} C = C \begin{array}{c}X\\ \diagup\\ \diagdown\\ Z\end{array}$$

Of particular practical importance are the following compounds:

Vinyl compounds
(for instance, styrene)

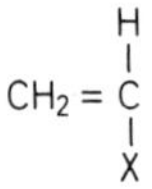

$$CH_2 = \underset{\underset{X}{|}}{\overset{\overset{H}{|}}{C}}$$

Vinylidene compounds
(for instance, methyl methacrylate);

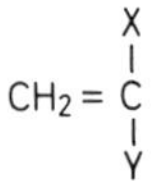

$$CH_2 = \underset{\underset{Y}{|}}{\overset{\overset{X}{|}}{C}} \quad \text{and}$$

1,2-disubstituted ethylene compounds
(for instance, maleic acid in UP resins):

$$X - CH = CH - Y$$

At room temperature, the monomers are stable compounds, but react readily with free radicals, thereby initiating a free radical chain reaction. Therefore, formation of free radicals reduces the stability in storage. Free radicals (molecules or molecule fragments with a high-energy unpaired electron) must be available for controlled initiation of a polymerization. The symbol $R \cdot$ is used to indicate these radicals, where the dot indicates the unpaired electron.

Every polymerization starts with the decomposition of the radical generator,

$$\underset{\text{Radical generator}}{I} \longrightarrow \underset{\text{Radical}}{2\,R\cdot} \tag{1}$$

This first step influences the kinetics of the polymerization but is in reality only a preliminary reaction.

Polymerization Reaction

Radical $R \cdot$ spontaneously forms an addition product with a monomer molecule:

Start of the chain reaction:

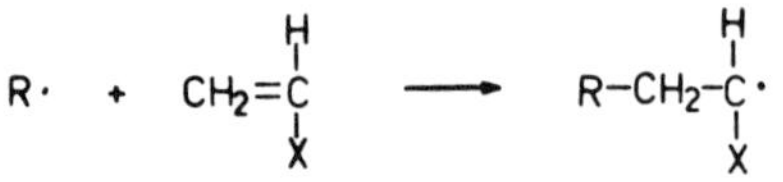

$$\underset{\text{Radical} + \text{Monomer}}{R\cdot \; + \; CH_2 = \underset{\underset{X}{|}}{\overset{\overset{H}{|}}{C}}} \longrightarrow \underset{\text{Monomer radical}}{R - CH_2 - \underset{\underset{X}{|}}{\overset{\overset{H}{|}}{C}}\cdot} \tag{2}$$

The addition does not destroy the free radical state; a new free radical molecule is being formed, which also can form an addition product with another or a second monomer molecule.

Chain reaction:

$$R-CH_2-\overset{\overset{\displaystyle H}{|}}{\underset{\underset{\displaystyle X}{|}}{C}}\cdot \quad + \quad CH_2=\overset{\overset{\displaystyle H}{|}}{\underset{\underset{\displaystyle X}{|}}{C}} \quad \longrightarrow \quad R-CH_2-\overset{\overset{\displaystyle H}{|}}{\underset{\underset{\displaystyle X}{|}}{C}}-CH_2-\overset{\overset{\displaystyle H}{|}}{\underset{\underset{\displaystyle X}{|}}{C}}\cdot \tag{3}$$

Monomer   +   Monomer   Chain radical
radical

In each step the growing chain $M_n\cdot$ is extended by one monomer unit M.

$$M_n\cdot + M \longrightarrow M_{n+1}^{\cdot} \tag{4}$$

The position of the free radical is always transferred to the last incorporated monomer.

An ideal chain reaction will last as long as there are monomer molecules available. Unavoidable interference from radical-trapping impurities (container walls belong in that category also) or due to the inherent instability of the free radical state are a characteristic feature of all chain reaction polymerizations.

Substances which are important radical scavengers are:

– Primary radicals
– Polymer radicals
– Chain transfer agents

Termination by chain transfer:

$$R\text{+}CH_2\text{-}\underset{\underset{\displaystyle X}{|}}{CH}\text{+}_n\cdot \quad + \quad H-Z \quad \longrightarrow \quad R\text{+}CH_2\text{-}\underset{\underset{\displaystyle X}{|}}{CH}\text{-}\text{)}_n\text{H} \quad + \quad Z\cdot \tag{5}$$

Polymer radical   +   Chain transfer   Unreactive   +   Radical
                          agent   molecule

The chain transfer agent $H-Z$ reacts with the radical center of the growing chain, and a terminated (dead) polymer chain is formed which is not influenced by the continuing polymerization reaction (not considering another radical transfer). The radical transfer agent $H-Z$ is transformed into a new, active radical center which is able to start a new chain reaction, similar to the chain initiator R. Chain transfer agent properties are exhibited by many components. These compounds may be purposely added to industrial polymerizations. During polymerization which is combined with a forming process, additives or other components present (for instance, fillers) may act as chain transfer agents and therefore terminate the chain reaction. This will result in a polymer with low molecular weight.

During polymerization of styrene the preferred termination of the chain reaction is brought about by combining the two chain molecules:

$$R\text{+}CH_2\text{-}\underset{\underset{\displaystyle X}{|}}{CH}\text{+}_n\cdot \quad + \quad \cdot\text{(}\underset{\underset{\displaystyle X}{|}}{CH}\text{-}CH_2\text{)}_m\text{R} \quad \longrightarrow \quad R\text{+}CH_2\text{-}\underset{\underset{\displaystyle X}{|}}{CH}\text{)}_{n+m}\text{R} \tag{6}$$

A large amount of available initiator radicals can also terminate the chain reaction:

$$R\text{+}CH_2\text{-}\underset{\underset{\displaystyle X}{|}}{CH}\text{)}_n\cdot \quad + \quad R\cdot \quad \longrightarrow \quad R\text{+}CH_2\text{-}\underset{\underset{\displaystyle X}{|}}{CH}\text{)}_n\text{-}R \tag{7}$$

Other types of chain termination reactions are caused by the instability of the chain radicals, which continue to react either spontaneously by elimination reaction or by combining two free radicals under disproportionation.

Chain termination by elimination:

$$R\text{-}(CH_2\text{-}CH\text{-})_n\text{-}CH_2\text{-}CH \cdot \quad \longrightarrow \quad R\text{-}(CH_2\text{-}CH\text{-})_n\text{-}CH\text{=}CH \quad + \quad H \cdot \qquad (8)$$

Chain termination by rearrangement (disproportionation):

$$R\text{-}(CH_2\text{-}CH\text{-})_n CH_2\text{-}CH \cdot \quad + \quad \cdot CH\text{-}CH_2\text{-}(CH_2\text{-}CH\text{-})_m R \quad \longrightarrow \qquad (9)$$

$$R\text{-}(CH_2\text{-}CH\text{-})_n\text{-}CH\text{=}CH \quad + \quad CH_2\text{-}CH_2\text{-}(CH_2\text{-}CH\text{-})_m R$$

6.1.2 Kinetics of Free Radical Polymerizations

Polymerization consists of an initiation reaction, a propagation reaction, and a chain termination reaction. During the reaction an equilibrium exists between the number of chain initiations and the number of chain terminations. In isothermal reactions a constant concentration of free radicals is established after a certain initial phase. These considerations lead to the following description of the polymerization reactions.

Formation of the radical:

$$I \longrightarrow 2\,R \cdot \qquad (10)$$

$I =$ initiator
$R \cdot =$ primary free radical

can be described within a low conversion rate by the equation:

$$\frac{d\,[R\cdot]}{dt} = v_0 = 2 \cdot k_0 \cdot [I] \qquad (11)$$

$v_0 =$ rate of free radical formation
$k_0 =$ decay constant of the initiator
$[\ ] =$ symbol for concentration

k_0 is specific for every free radical initiator; therefore, the starting rate of a polymerization is dependent on the initiator. The primary radical that is formed is added to the monomer molecules:

$$v_1' = k_1' \cdot [M] \cdot [R\cdot] \qquad (12)$$

$v_1' =$ rate of primary step
$k_1' =$ constant of addition of primary radicals on to monomer molecules
$M =$ monomer

For the initial step of the reaction one usually combines Equations (11) and (12). One has to consider that only a certain fraction, designated by the factor f, of a specific initiator is active and not the total theoretical amount of primary free radicals.

$$v_1 = f \cdot k_o \cdot k_1' \cdot [M] \cdot [I] \tag{13}$$

$$k_1 = f \cdot k_o \cdot k_1' \tag{13a}$$

$$v_1 = k_1 \cdot [M] \cdot [I] \tag{13b}$$

$v_1 =$ initial rate
$k_1 =$ initial rate constant

Chain propagation is determined by a specific chain propagation, the concentration of the monomer radical, and the concentrations of chain radical and monomer.

$$v_2 = k_2 \cdot [M_x \cdot] \cdot [M] \tag{14}$$

$v_2 =$ rate of chain propagation
$k_2 =$ propagation constant
$M_x \cdot =$ free radical molecule with x units of monomer

The monomer consuming reactions compete with the chain-termination reactions. The two most important termination reactions are combination and rearrangement. Both reactions are bimolecular terminations which can be described according to the following equation:

$$v_3 = -\frac{d\,[M_x \cdot]}{dt} = k_3 \cdot [M_x \cdot]^2 \tag{15}$$

$v_3 =$ chain termination rate
$k_3 =$ chain cleavage rate constant

Chain termination caused by primary free radicals is of importance to the polymerization process carried out in the processing plant. The following equation describes the chain termination during the curing process of unsaturated polyester (UP) resins with high initiator concentration and at high temperature:

$$v_3' = -\frac{d\,[M_x \cdot]}{dt} = k_3' \cdot [M_x \cdot] \cdot [R \cdot] \tag{16}$$

During polymerization in the processing plant, reactions with contaminants may cause chain termination. A quantitative evaluation of these influences is extremely difficult.

If one restricts chain termination to combination and disproportionation (that is, without considering free radical chain transfer), an equilibrium is reached between the rate of these reactions and the rate of initiation follows:

$$v_1 = v_3 \tag{17}$$

$$v_1 = k_3 \cdot [M_x \cdot]^2$$

According to this equation one can calculate the stationary concentration of free radicals:

$$[M_x \cdot] = \left(\frac{v_1}{k_3}\right)^{\frac{1}{2}} \tag{18}$$

If one inserts Equation (18) into Equation (14) a general equation for the rate of propagation is obtained:

$$v_2 = -\frac{d\,[M]}{dt} = k_2 \cdot [M_x \cdot] \cdot [M] = k_2 \cdot \left(\frac{v_1}{k_3}\right)^{\frac{1}{2}} [M] \tag{19}$$

The rate of polymerization is dependent on:

(a) the rate constant for the chain propagation k_2, which is specific for the monomer and is dependent on the temperature according to an exponential function (Arrhenius):

(b) the rate of initiation v_1, which itself is dependent on the type of initiator and the temperature as well as the concentration of the free radical generator and the monomers;

(c) the rate of chain termination, which is specific to the type of monomer and dependent of the temperature, kinematic viscosity, and several contaminations; and finally

(d) the monomer concentration, which changes constantly during polymerization.

6.1.3   Molecular Weight and Molecular Weight Control

The interaction between chain propagation (v_2) and chain termination (v_3) determines the molecular weights of the forming polymers. The propagation probability, is the quotient of the propagation reaction and the sum of all reaction rates (with the exception of the initiation reaction). When only bimolecular termination reactions are considered, the propagation probability is:

$$\gamma = \frac{v_2}{v_2 + v_3} = \frac{k_2 \cdot [M]}{k_2 \cdot [M] + k_3 \cdot \left(\frac{v_1}{k_3}\right)^{\frac{1}{2}}} \tag{20}$$

High molecular weights are the result of a large value of γ. Therefore the rate constant of the chain propagation has a positive influence on the molecular weight, while the rate constant of the chain termination (reaction) and the rate of the chain initiation reaction have a negative influence. Consequently, two possibilities exist to influence the size of the molecular weights:

(a) Increase the concentration of initiator radicals; that means if one uses a high concentration of initiator, v_1 increases and the resulting molecular weights decrease.

(b) The use of high reaction temperatures favors chain termination, resulting in lower molecular weights.

6.1.4   Rate of Polymerization and Its Control

According to Equation (19), the rate of reaction is affected by:

v_1,   and by the type of initiator and the temperature

k_2,   a monomer-specific and temperature-dependent constant

k_3,   one or more chain termination constants which are also monomer and temperature-dependent, and

[M]   the monomer concentration.

Since polymerization which occurs in the processing plant is always mass polymerization, it is not possible for the processor to modify [M]. To vary the type of monomer in order to modify the rate of reaction is hardly worth considering. This leaves as variables the type of initiator and the temperature. In order to get high rates of polymerization one has to use high concentrations of free radicals and high temperatures.

An increased rate of polymerization will shorten the reaction time, but the factors which increase the rate of polymerization reduce the size of the molecular weight at the same time. Therefore, shorter reaction time results in poorer quality of the molding compound. In many cases a high rate of polymerization is also undesirable from a point of view.

6.1.5   Copolymerization

Usually, polymerizations of two or more monomers are carried out at the same time in the polymer processing plant. The curing of UP resins is a free- radical-induced copolymerization of styrene and fumaric acid diesters. During the reaction casting of PMMA, co-monomers can be added as plasticizers or cross-linking compounds.

In principle, copolymerization takes place according to the laws of homopolymerization, but specific reactivities of each different monomer and chain cause the copolymerization of two monomers to be characterized by four different growth reactions. If one labels the two monomers M_1 and M_2 and the resulting chain radicals $-M_1\cdot$ and $-M_2\cdot$, the following reactions are possible:

$$R - M_1\cdot + M_1 \longrightarrow R - M_1 - M_1\cdot \tag{21}$$

$$R - M_1\cdot + M_2 \longrightarrow R - M_1 - M_2\cdot \tag{22}$$

$$R - M_2\cdot + M_1 \longrightarrow R - M_2 - M_1\cdot \tag{23}$$

$$R - M_2\cdot + M_2 \longrightarrow R - M_2 - M_2\cdot \tag{24}$$

In reactions (21)–(24) the rates of reaction and the rate constants are marked with two-digit indices; the first digit identifies the radical monomer and the second digit the addition monomer molecule.

$$v_{11} = k_{11}\cdot[M_1\cdot]\cdot[M_1] \tag{25}$$

$$v_{12} = k_{12}\cdot[M_1\cdot]\cdot[M_2] \tag{26}$$

$$v_{21} = k_{21}\cdot[M_2\cdot]\cdot[M_1] \tag{27}$$

$$v_{22} = k_{22}\cdot[M_2\cdot]\cdot[M_2] \tag{28}$$

Some monomer pairs are very easily copolymerized, and there are others which do not copolymerize at all. Homopolymerization is more likely to occur if the rates of reaction v_{12} and v_{21} are very small compared to v_{11} and v_{22}. The monomer reactivity ratios r_1 and r_2 are measures of the behavior of two monomers during copolymerization. The quotient r_1 consists of the rate constant for adding M_1 to $-M_1\cdot$ and the rate constant for adding M_2 to $-M_1\cdot$:

$$r_1 = \frac{k_{11}}{k_{12}} \tag{29}$$

accordingly, r_2 is defined for the addition to $-M_2\cdot$:

$$r_2 = \frac{k_{22}}{k_{21}} \tag{30}$$

If the monomer reactivity ratios are known, then one can predict if a copolymerization is possible and which ratio of the two monomers will result in a polymer with a composition that corresponds to the monomer composition. For small monomer conversions, the ratio (number of moles) p of M_1 to M_2 in a polymer is given by:

$$p = \frac{M_1}{M_2} = \frac{v_{11}+v_{21}}{v_{12}+v_{22}} = \frac{k_{11}\cdot[M_1\cdot]\cdot[M_1]+k_{21}\cdot[M_2\cdot]\cdot[M_1]}{k_{12}\cdot[M_1\cdot]\cdot[M_2]+k_{22}\cdot[M_2\cdot]\cdot[M_2]} \tag{31}$$

Equation (31) is valid only at low temperatures.

If one assumes that in order to maintain a steady concentration of radicals:

$$v_{12} = v_{21}$$

one can rewrite equation (31):

$$p = \frac{\dfrac{v_{11}}{v_{12}}+1}{\dfrac{v_{22}}{v_{21}}+1} = \frac{\dfrac{k_{11}\cdot[M_1]}{k_{12}\cdot[M_2]}+1}{\dfrac{k_{22}\cdot[M_2]}{k_{21}\cdot[M_1]}+1} \tag{32}$$

or

$$p = \frac{r_1 m+1}{\dfrac{r_2}{m}+1} \qquad \text{for } m = \frac{[M_1]}{[M_2]} \tag{33}$$

M equals 1 if equimolar amounts of monomer are being used; then the composition of the product can be calculated according to:

$$p = \frac{r_1+1}{r_2+1} \qquad \text{(for } m=1) \tag{34}$$

To find the molar ratio $[M_1]/[M_2]$, where the composition of the product corresponds to the composition of the monomer, p has to equal m; (p=m). Equation (33) can then be written as:

$$m = \frac{r_2-1}{r_1-1} \tag{35}$$

6.1.6   The Gel Effect

The above-mentioned kinetic equations are valid only at low conversions or for dilute solutions. At higher conversions the reaction kinetic is different, because the rate of reaction is determined by diffusion processes. The gel effect appears in polymerizations in which the chain-termination reaction is a bimolecular reaction. The rate of polymerization increases automatically if the mobility of the chain radicals decreases and if the diffusion of the chain radicals is rate-determining for the termination reaction. The reason for this is that fewer free radicals are consumed by termination reactions, which leads to formation of larger molecules. Finally, at very high conversions the addition of monomer is controlled by diffusion.

The gel effect is especially strong during polymerization of methyl methacrylate and will be described in detail in Section 6.3.2

Bibliography to Section 6.1

[1] *Batzer, H.:* »Einführung in die makromolekulare Chemie«, Alfred Hüthig Verlag, Heidelberg, 1973.

[2] *Bikales, N. M.:* »Encyclopedia of Polymer Science and Technology Plastics, Resins, Rubbers, Fibers«, Volume 1–15, Interscience Publishers Inc., New York-London-Sydney-Toronto, 1964 bis 1971.

[3] *Braun, D., H. Cherdron, W. Kern:* »Praktikum der makromolekularen Chemie«. Alfred Hüthig Verlag, Heidelberg, 1971.

[4] *Cowie, J.:* »Chemie und Physik der Polymeren«. Verlag Chemie, Weinheim, 1976.

[5] *Elias, H. G.:* »Makromoleküle«. 3. Aufl., Alfred Hüthig Verlag, Heidelberg, 1973.

[6] *Ham, G. E.:* »Kinetics and Mechanisms of Polymerisation Series«, Bd. 1. Marcel Dekker, Inc., New York, 1967.

[7] *Haman, K.:* »Chemie der Kunststoffe«, Sammlung Göschen 1173/1173a. Verlag W. de Gruyter & Co., Berlin 1967.

[8] *Henrici-Olivé, G., S. Olivé:* »Polymerisationen«. Verlag Chemie GmbH., Weinheim, 1969.

[9] *Houwink, R., A. J. Staverman:* »Chemie und Technologie der Kunststoffe«, Bd. 1. Akademische Verlagsgesellschaft, Leipzig, 1962.

[10] *Houwink, R.:* »Elastomers and Plastomers, Their Chemistry, Physics and Technology«, Bd. 1. Elsevier Publishing Co., Inc., New York-Amsterdam-London-Brüssel, 1950.

[11] *v. Meysenburg, C. M.:* »Kunststoffkunde für Ingenieure«. Carl Hanser Verlag, München, 1968.

[12] *Patfoort, G. A.:* »Polymers«. E. Story-Scientia, Gent, 1974.

[13] *Schreiber, J.:* »Chemie und Technologie der künstlichen Harze«, Bd. 1. Wissenschaftliche Verlagsgesellschaft m. b. H., Stuttgart, 1961.

[14] *Schulz, G.:* »Die Kunststoffe«. Carl Hanser Verlag, München, 1959.

[15] *Vollmert, B.:* »Grundlagen der makromolekularen Chemie«. Springer Verlag, Berlin-Göttingen-Heidelberg, 1962.

6.2   Free Radicals as Processing Aids

Free radical generators are necessary for curing reactions which are based on radical polymerization. Decomposition of the radical generator initiates the curing reaction; the decomposition has to occur at a time predetermined by the processor. Premature decomposition can cause severe processing difficulties; the economics of the process are affected adversely if decomposition occurs too late. Free radicals are not only used to initiate radical polymerization, they are also required for initiation of cross-linking reactions.

Free radical generators are processing aids which will start the reaction. Their use is subject to special conditions, and the radical generators and processing methods must be carefully matched. The various processes and curing conditions of free radical generators must be taken into account.

Table 7   Processes using radical polymerization or a radical crosslinking reaction [1]

	Starting material	Symbol
Polymerization	Methyl methacrylate	PMMA
	UP resins	UP
	Diallylesters	
Cross-linking reaction	Natural rubber	NR
	Isoprene rubber	IR
	Butadiene rubber	BR
	Chloroprene rubber	CR
	Butadiene rubber	SBR
	Acrylonitrile rubber	NBR
	Methyl silicone rubber	SI
	Urethane rubber	
	Acrylic ester-Acrylonitrile Copolymers	ANM
	Ethylene-propylene rubber	EPM
	Polysulfide rubber	TR
	Polyethylene HD and LD	PE
	Chloropolyethylene	CM/CPE
	Polypentene-1	PB
	Ethylene-vinyl acetate Copolymers	EVA
	Acrylonitrile-butadiene-styrol Terpolymers	ABS
	Polyvinyl chloride	PVC
	Polyisobutylene	IM

Free radicals can be generated in several different ways:
(a) By thermally exciting certain monomer molecules; for example, styrene
(b) By treating the plastic or auxiliary component with high-energy radiation
(c) By thermal decomposition of unstable components
(d) By photolytic decomposition of photoinitiators
(e) By spontaneous decomposition of unstable components in an ultrahigh frequency field.

Thermally excited styrene forms a dimer, which reacts with a third styrene molecule to form a styryl radical. The activation energy necessary for this reaction is already available at room temperature. Therefore, styrene, UP resins, and other styrene-containing solutions have only limited stability, and stabilizers have to be added to avoid premature reactions.

High-energy radiation such as gamma radiation or high-energy electrons interact with an active resin or a polymer and thus generate free radicals by the cleavage of covalent bonds. This process can be controlled by the use of unstable additives and allows polymerizations at very low temperatures. In the polymer processing plant this form of free radical generation is used to cross-link PE by high energy electrons.

6.2.1 Radical Generation by Thermal Decomposition of Radical Initiators

Thermal decomposition of unstable components is the most widely used method for generating primary free radicals.

Unstable components which are currently used include

- Peroxides
- Azo compounds
- Radical generators without peroxy or azo groups

The technical application of radical generators without peroxy or azo groups has just begun. Compounds that belong to this class include dibenzyl compounds and benzpinacol.

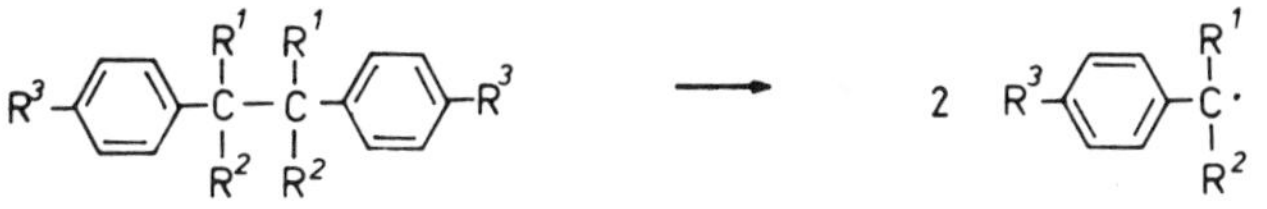

Dibenzyl compound

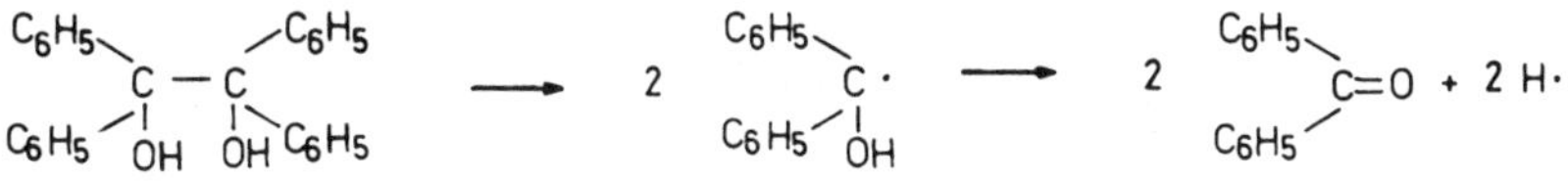

Benzpinacol

These radical generators are insensitive to shock and are easier to handle than the peroxides. No gaseous compounds are released, which is a distinct advantage compared to the azo compounds.

Only a few azo compounds are being used as radical generators (Table 8), while a wide variety of peroxides are employed. In the polymer processing plant, azo compounds are used during reaction casting of PMMA and during polymerization of diethylene glycol-

Table 8 Azo compounds as auxiliary compounds for the generation of radicals

Name	Formula	Half lives [°C]			
		30°	50°	70°	90°
2,2′-Azo-bis (2,4-Dimethyl valero nitril)	$(CH_3)_2CH-CH_2-C(CH_3)(CN)-N=N-C(CH_3)(CN)-CH_2-CH(CH_3)_2$	29 d	29 h	1,5 h	4 min
2,2′-Azo-bis (Iso butyric nitrile) (AIBN)	$NC-C(CH_3)_2-N=N-C(CH_3)_2-CN$		6,3 d	7,5 h	25 min

bis(allyl carbonate). Two free radicals and gaseous N_2 are formed during the decomposition of azo compounds:

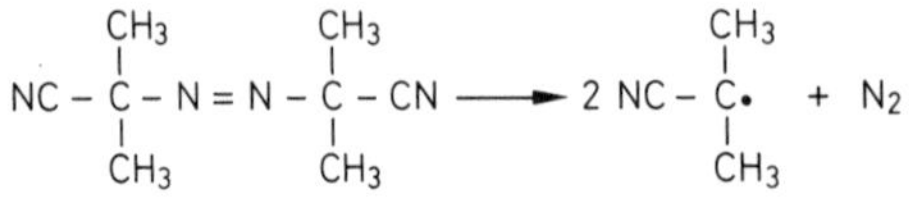

$$
\underset{\underset{CH_3}{|}}{\overset{\overset{CH_3}{|}}{NC - C}} - N = N - \underset{\underset{CH_3}{|}}{\overset{\overset{CH_3}{|}}{C}} - CN \longrightarrow 2 \; NC - \underset{\underset{CH_3}{|}}{\overset{\overset{CH_3}{|}}{C\bullet}} \; + \; N_2
$$

Decomposition of 2,2 azobis [isobutyric nitrile] AIBN

Peroxides are the most important radical generators. Peroxides are a group of compounds which are unstable and contain at least one $-O-O-$ group (peroxide group). In Table 10 one will find different classes of peroxides according to the structure of the remaining molecule.

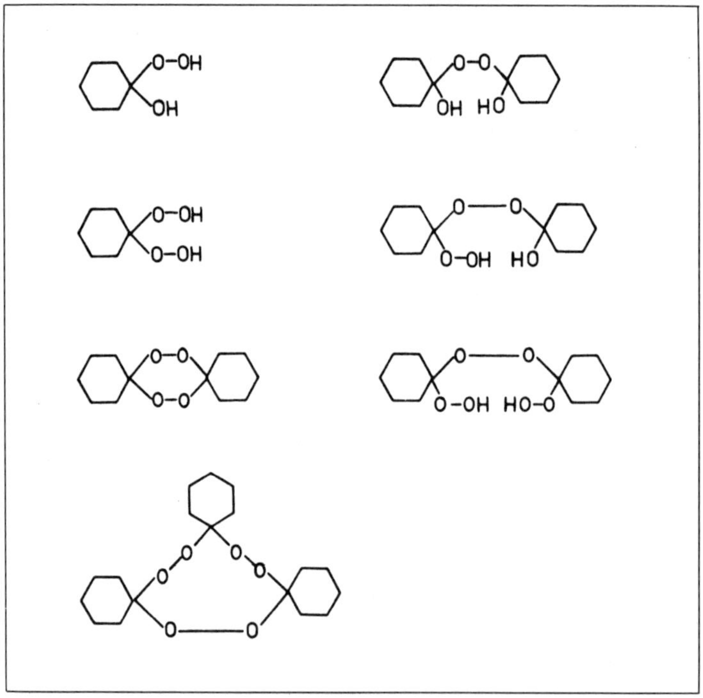

A multitude of structures are possible even within the individual peroxide classes [2]. Seven different ketone peroxides alone can be obtained by reacting cyclohexanone with hydrogen peroxide (Table 9). A number of peroxides are offered in different formulations which are a combination of pure peroxide and various additives, such as filler, plasticizer, solvent, or water. In some cases, additives are used to produce a more suitable compound; for example an inert compound is added to make the peroxide less sensitive to shock.

Table 10   Overview of types of peroxides

Types of Peroxides	General Formula
Percarboxylic acid	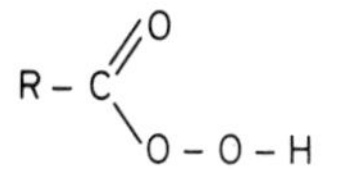 $R - C{\overset{\displaystyle O}{\underset{\displaystyle O-O-H}{<}}}$
Diacylperoxide	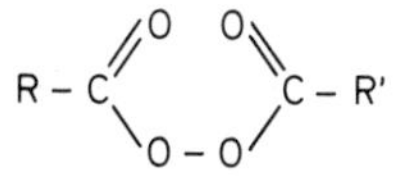 $R - C{<} \overset{O \quad O}{\underset{O-O}{}} {>} C - R'$
Percarboxylate	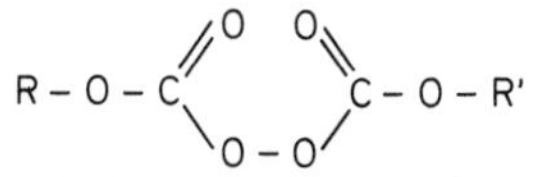 $R - O - C{<}\overset{O \quad O}{\underset{O-O}{}}{>}C - O - R'$
Acetylalkylsulfonyl peroxide	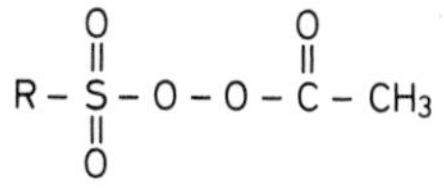 $R - \overset{O}{\underset{O}{\overset{\|}{\underset{\|}{S}}}} - O - O - \overset{O}{\overset{\|}{C}} - CH_3$
Perester	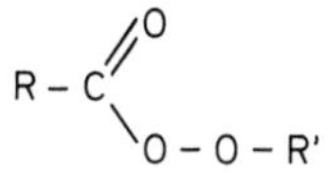 $R - C{\overset{\displaystyle O}{\underset{\displaystyle O-O-R'}{<}}}$
Perketal	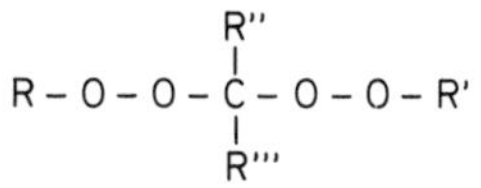 $R - O - O - \overset{R''}{\underset{R'''}{\overset{\|}{\underset{\|}{C}}}} - O - O - R'$
Diaryl- Alkyl-aralkyl- } peroxide Dialkyl-	$R - O - O - R'$
Hydroperoxide	$R - O - OH$
Ketone peroxide	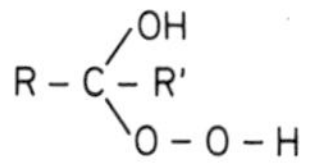 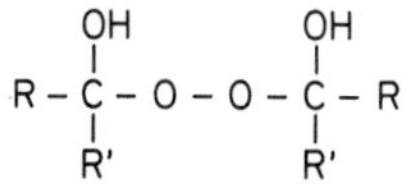 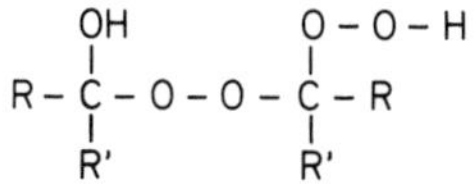 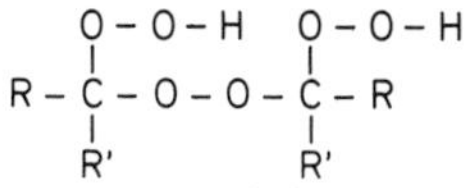 $R - C{\overset{\displaystyle OH}{\underset{\displaystyle O-O-H}{<}}} - R'$ $R - \overset{OH}{\underset{R'}{\overset{\|}{\underset{\|}{C}}}} - O - O - \overset{OH}{\underset{R'}{\overset{\|}{\underset{\|}{C}}}} - R$ $R - \overset{OH}{\underset{R'}{\overset{\|}{\underset{\|}{C}}}} - O - O - \overset{O-O-H}{\underset{R'}{\overset{\|}{\underset{\|}{C}}}} - R$ $R - \overset{O-O-H}{\underset{R'}{\overset{\|}{\underset{\|}{C}}}} - O - O - \overset{O-O-H}{\underset{R'}{\overset{\|}{\underset{\|}{C}}}} - R$

Table 11   Examples of commercial peroxides

Name	Formula	Half-life at 100 °C	Field of application
Acetylcyclo-hexylsulfonyl-peroxide	$CH_3 - \overset{\overset{O}{\|\|}}{C} - O - O - \overset{\overset{O}{\|\|}}{\underset{\underset{O}{\|\|}}{S}} -$ (cyclohexyl)	6 s	PMMA reaction casting
Bis(4-t.-butyl. cyclohexyl)per-oxidicarbonate	$\left[CH_3 - \overset{\overset{CH_3}{\|}}{\underset{\underset{CH_3}{\|}}{C}} - (cyclohexyl) - O - \overset{\overset{O}{\|\|}}{C} - O - \right]_2$	0,5 min	Casting of diallyl-esters
Bis-2,4-dichlor-benzoyl-peroxide	$Cl-(ring)-\overset{\overset{O}{\|\|}}{C} - O - O - \overset{\overset{O}{\|\|}}{C}-(ring)-Cl$ (with Cl)	1,1 min	Crosslinking of silicone rubber
Di-Lauroyl-peroxide	$H + CH_2 \}_{11} - \overset{\overset{O}{\|\|}}{C} - O - O - \overset{\overset{O}{\|\|}}{C} + CH_2 \}_{11} - H$	5,5 min	PMMA reaction casting
tert.-Butyl-peroctoate	$CH_3 - \overset{\overset{CH_3}{\|}}{\underset{\underset{CH_3}{\|}}{C}} - O - O - \overset{\overset{O}{\|\|}}{C} - CH_2 - \overset{\underset{\underset{C_2H_5}{\|}}{}}{CH} - CH_2 - CH_2 - CH_3$	10 h	UP cold curing
tert.-Butyl-perbenzoate	$CH_3 - \overset{\overset{CH_3}{\|}}{\underset{\underset{CH_3}{\|}}{C}} - O - O - \overset{\overset{O}{\|\|}}{C} - (ring)$	20 h	UP cold curing UHF cross-linking
Di-Cumyl-peroxide	$(ring)-\overset{\overset{CH_3}{\|}}{\underset{\underset{CH_3}{\|}}{C}} - O - O - \overset{\overset{CH_3}{\|}}{\underset{\underset{CH_3}{\|}}{C}}-(ring)$	(6 min. 160 °C)	UP hot curing PE cross-linking
Di-tert.-Butylperoxide	$CH_3 - \overset{\overset{CH_3}{\|}}{\underset{\underset{CH_3}{\|}}{C}} - O - O - \overset{\overset{CH_3}{\|}}{\underset{\underset{CH_3}{\|}}{C}} - CH_3$	(20 min, 160 °C)	UP hot curing PE cross-linking
Cumolhydro-peroxide	$(ring)-\overset{\overset{CH_3}{\|}}{\underset{\underset{CH_3}{\|}}{C}} - O - OH$	(9 h, 160 °C)	UP hot curing
tert.-Butyl-hydroperoxide	$CH_3 - \overset{\overset{CH_3}{\|}}{\underset{\underset{CH_3}{\|}}{C}} - O - OH$	(30 h, 160 °C)	UP hot curing

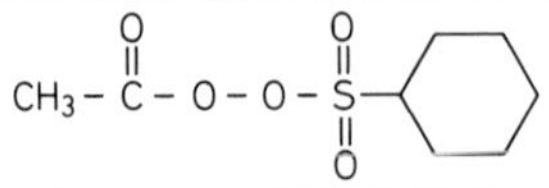

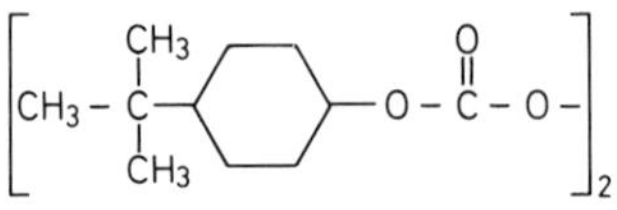

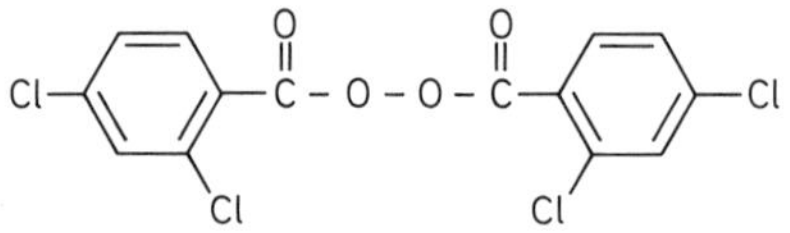

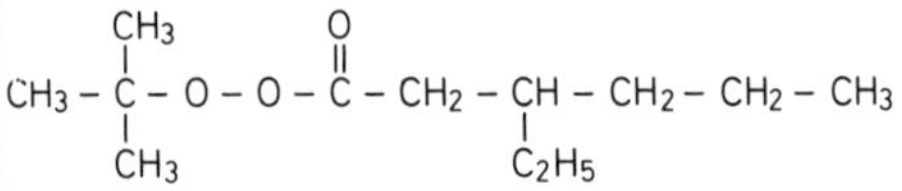

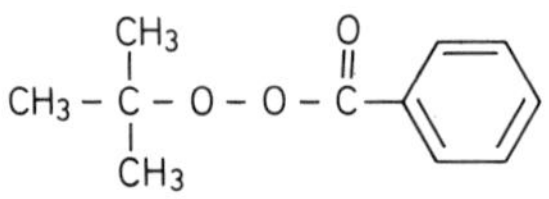

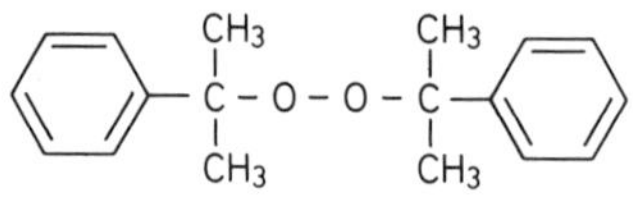

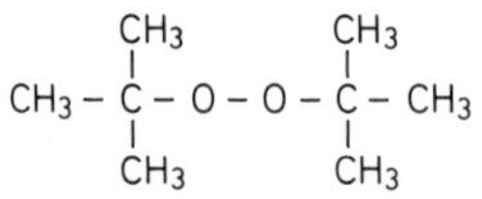

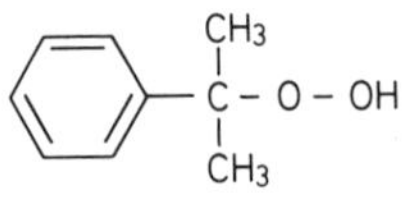

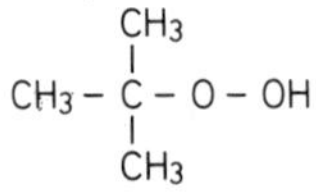

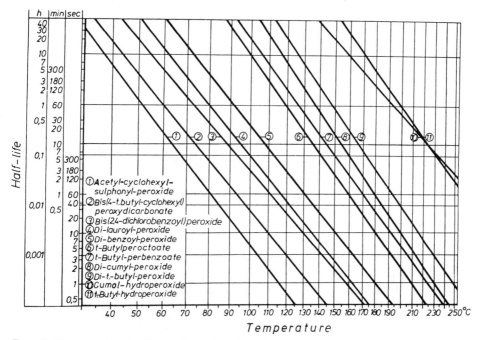

Figure 2   Decomposition half-lives of peroxides dependent on temperature [3]

Peroxides are cleaved into free radicals by thermal energy:

$$R-O-O-R \quad \xrightarrow{\text{Heat}} \quad 2\ R-O\bullet$$

The remainder R determines the bonding energy of the $-O-O-$ bond and therefore the thermal stability of the peroxides. Peroxides with differently structured unit R's show different decomposition temperatures (see Figure 2). Secondary products formed by decomposition of the radicals are also determined by the unit R. Three examples will demonstrate the different secondary reactions.

(a) Decomposition of dibenzoyl peroxide:

(c) Decomposition of tert-butyl perbenzoate:

The most important applications, namely, polymerization and cross-linking, require radicals with different reactivities. Polymerization is initiated by adding a radical to an olefinic double bond. Even low energy radicals can make this addition reaction occur. Theoretically, then, most peroxides are suitable polymerization initiators. The effectiveness of the radical in various cross-linking reactions depends on its ability to abstract an H atom from the compound to be cross-linked. Therefore, the substrate acts as a chain-transfer agent. The greater the energy drop between the primary radical and the polymer radical, the more efficient this type of radical transfer becomes. CH_3 (methyl) radicals are preferred agents for cross-linking reactions.

Thermal decomposition of the peroxide is usually a first-order reaction. Radicals already generated are able to induce decomposition of the peroxide (inductive decomposition). The amount of radicals formed per unit time is directly proportional to the peroxide concentration and the temperature. (See Sec. 6.12)

Not every one of the generated radicals is useful for initiation of a polymerization or cross-linking reaction. A "cage effect" can cause a reaction between the radicals or their decomposition products before the primary radicals are able to react with the intended compound. Through secondary reactions or a combination of radicals, this cage effect produces inactive products or more stable radicals. Therefore only a portion of the primary radicals are available for the desired reaction.

According to the following equation, one can calculate the amount of peroxide decayed after a certain time and at constant temperature:

$$C = C_0 \cdot e^{-k \cdot t}$$

C = amount of peroxide decayed
C_0 = amount of peroxide at t equals zero
k = decay rate constant
t = reaction time

Peroxides are usually characterized by their half-life at different temperatures. If $C = 0.5 \, C_0$ is substituted into the equation above, the following equation for the half-life results:

$$t_{1/2} = \frac{\ln 2}{k}$$

The half-life is independent of the concentration but is valid only for certain temperatures.

The graph in Figure 3 shows clearly that within a certain narrow temperature range, a rise in temperature results in a sharp decrease in the half-life.

Peroxides can be used only within this narrow temperature range. Decomposition of the peroxide is exothermic.

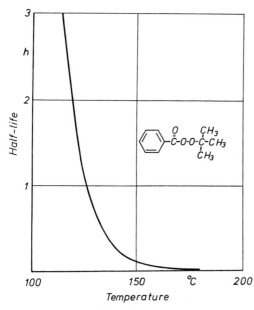

Figure 3
Half-life of tert-butyl perbenzoate
as a function of temperature
(linear scale).

The use of additives to accelerate decomposition of radical generators containing per-oxide groups is of particular technical importance, especially for the cold setting of UP resins. Two different mechanisms may be responsible for increasing the rate of decomposition:

(a) catalytic decomposition of the peroxide by low oxidation number metal ions, or
(b) decomposition by reduction of oxidizable components.

In the polymer processing plant, the term "accelerator" is used for compounds which act as catalysts as well as for those which act as reducing agents.

Salts of cobalt and vanadium are particularly effective accelerators that act as catalysts. Metal salts of long-chain carboxylic acids such as octane carboxylic acid or naphthenic acid are soluble in UP resins and other polymerizable compounds.

The decomposition of ketone peroxides and hydroperoxides is most effective when co-balt accelerators are used. During the curing process with cobalt salts, intermediate oxidation of two-valent cobalt to the three-valent cobalt occurs, followed by peroxide induced reduction to the two-valent state.

$$CH_3-\underset{\underset{CH_3}{|}}{\overset{\overset{CH_3}{|}}{C}}-O-OH \; + \; Co^{++} \longrightarrow CH_3-\underset{\underset{CH_3}{|}}{\overset{\overset{CH_3}{|}}{C}}-O\bullet \; + \; Co^{+++} \; + \; OH^-$$

$$CH_3-\underset{\underset{CH_3}{|}}{\overset{\overset{CH_3}{|}}{C}}-O-OH \; + \; Co^{+++} \longrightarrow CH_3-\underset{\underset{CH_3}{|}}{\overset{\overset{CH_3}{|}}{C}}-O-O\bullet \; + \; Co^{++} \; + \; H^+$$

The following secondary reaction may occur:

$$CH_3-\underset{\underset{CH_3}{|}}{\overset{\overset{CH_3}{|}}{C}}-O\bullet \; + \; Co^{++} \longrightarrow CH_3-\underset{\underset{CH_3}{|}}{\overset{\overset{CH_3}{|}}{C}}-O^- \; + \; Co^{+++}$$

The use of large amounts of the metal salts promotes alcoholate formation [2]; therefore, only catalytic amounts of metal ions should be used as accelerators for UP resin curing. The catalyst is never consumed since the catalytically active two-valent ion is constantly regenerated. Strictly thermal decomposition of the hydroperoxide results in two radicals; on the other hand, decomposition induced by metal catalysis delivers only 1 mole of radical per mole of peroxide.

Amines are particularly effective in accelerating the decomposition of diacyl peroxides. Initially the attack of the nucleophilic amino group on the electrophilic peroxide group produces an unstable complex, which decomposes very rapidly. The greater the nucleophilic character of the amino group, the greater is the accelerating effect unless the approach of the reactants is hindered sterically.

All of the generated radicals are able to initiate the polymerization and will be incorporated, in one form or another, in the final product. It follows that amines are not true catalysts, but reactants which are consumed during the reaction. Therefore, a different ratio of the quantity of peroxide to accelerator is required in these cases; generally, the ratio lies between 1 and 0.1-0.8, with the ratio of 1:0.4 used most commonly [4].

6.2.2   Photolytic Decomposition of Radical Initiators

Certain compounds, which belong to various classes of compounds (Table 12), decompose into free radicals under the influence of light. This process has a distinct advantage in that it does not depend on a certain minimum temperature (although elevated temperatures can enhance the formation of radicals). The disadvantage of the photolytic generation of radicals is that the light required for bond cleavage will also interact with the material to be processed; the result is that the intensity of the light decreases in proportion to the increasing depth of penetration.

Therefore, applications of the photolytic radical formation are quite limited. Cross-linking is impossible in a closed mold. Examples of applications are the curing of thin layers of coatings and filament-wound objects made from UP resins. Cross-linking of thin parts made from polyolefins is also conceivable.

Table 12   Free radical generators which can be cleaved photolytically

Functional Group	Example	Formula and decomposition
Carbonyl-compound	Benzoin ethylether	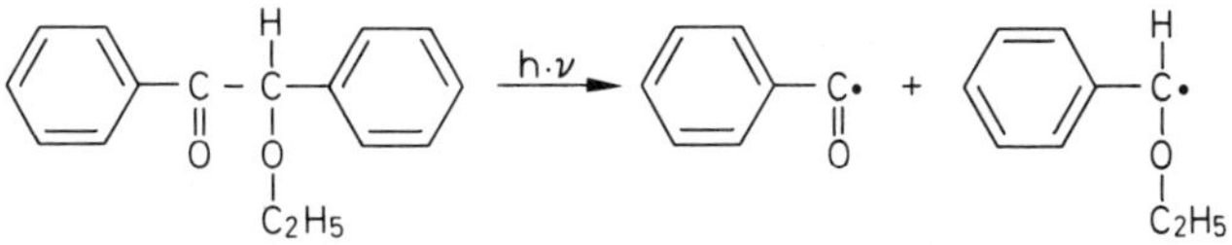
Organic sulfur compound	Dibutyl sulfide	$C_4H_9 - S - S - C_4H_9 \xrightarrow{h \cdot \nu} 2 \ C_4H_9 - S \cdot$
	Desyl phenyl sulfide	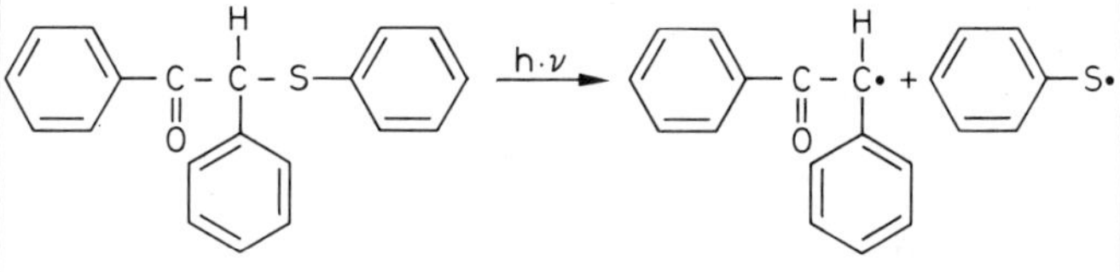
Peroxides	Di-tert-butyl peroxide	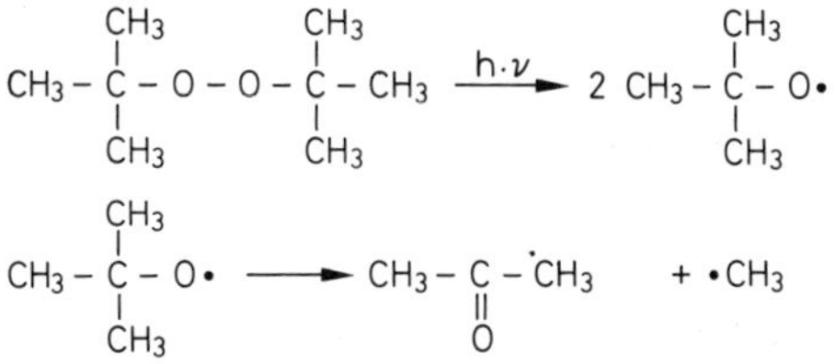
Azo- and diazo-compounds	Azobisisobutyric acid nitrile	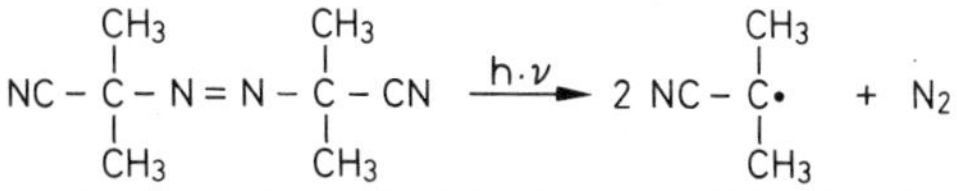
Halogen compounds	Carbon tetrachloride	$CCl_4 \xrightarrow{h \cdot \nu} Cl \cdot \ + \ Cl_3C \cdot$
	2-Naphthalene-sulfochloride	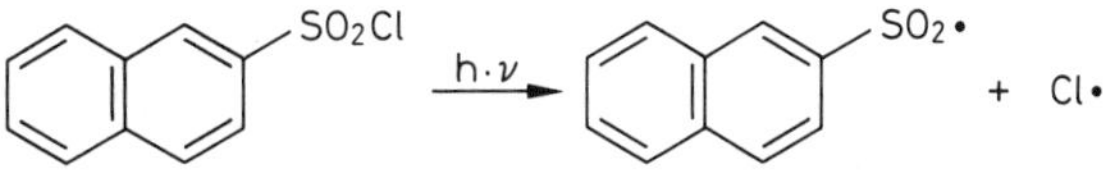
	Silver bromide	$AgBr \xrightarrow{h \cdot \nu} Ag \cdot \ + \ Br \cdot$

According to the Lambert-Beer law, the intensity of light decreases as the distance from the surface increases.

$$I_a = I_0 \cdot (1 - e^{-\varepsilon \cdot a \cdot c})$$

I_a = light intensity of depth of layer
I_0 = light intensity on the surface
ε = extinction coefficient
a = distance of point a from the surface
c = concentration of the component which has the extinction coefficient ε

The rate of formation of free radicals is directly proportional to the light quantum yield and the local light intensity:

$$v_R \approx \text{(H)} \cdot I_a$$

V_R = rate of radical generation
(H) = quantum yield of light

The light intensity decreases logarithmically as distance from the surface increases, and the same is true for the decrease in the rate of formation of radicals. Usually, prior to cross-linking, the polymer itself absorbs light. This causes its molecules to become excited, thus causing it to either heat up or decompose. Photoinitiators which show energy absorption maxima at wavelengths unaccompanied by interaction between light and polymer are ideal. Under optimum conditions the light source emits only those wavelengths which will be absorbed by the photoinitiator and not by the polymer. For these reasons, special demands are made on the photoinitiators. Furthermore, an application in the polymer processing plant requires sufficient thermal stability, adequate solubility in the reactant and high quantum yield. A table of photoinitiators can be found in [5]. Organic carbonyl compounds are particularly suitable photoinitiators. For practical applications, the following radical generators are used for curing of UP resins:

- Benzoin ethers
- Benzil dialkyl ketals

Besides benzoin ether, benzil dimethyl ketal is particularly useful because it forms very active methyl radicals during decomposition.

The maximum extinction at 300–400 nm coincides relatively well with the maximum emission of common mercury vapor lamps.

Thermally unstable radical generators, such as peroxides and azo compounds, also undergo photolytic decomposition (Table 12). Additives accelerate radical generation by photolysis.

6.2.3   Decomposition of Peroxides in a UHF Alternating Field

A polar polymer, or polar reaction resin, is heated if placed in a UHF field. Heat created in this field can be used to decompose a peroxide. This technique of free radical generation is independent of both thermal conduction and radiation. The rise in temperature is caused by the dielectric loss factor in the reactant itself. In principle, all peroxides can be decomposed according to this method.

Nonpolar polymers and nonpolar resins do not show a rise in temperature in a UHF field. But generation of free radicals is still possible if the peroxide has a polar molecular structure, allowing it to interact with the UHF field and absorb energy from it [6, 7].

In the UHF field, cross-linking of nonpolar polymers (such as PE) is possible only with UHF sensitive peroxides. The latter are not needed for polymerization of methyl methacrylate and diallyl esters, curing of UP resins, or cross-linking of polymers (e. g. chloro rubber, nitrile rubbers, PVC, and polymers containing carbon black) because the reactant itself absorbs energy from the UHF field and then transfers the energy to the peroxide.

2,4-dichloro benzoyl peroxide present in silicone oil decomposes very rapidly if the mixture is brought into a UHF field. The strongly polar C–Cl group of this peroxide causes such great dielectric losses that the temperature quickly rises to the point where rapid peroxide decomposition occurs. At 74 °C 2,4-dichloro benzoyl peroxide has a half-life of 1 hour; therefore it can be used as a cross-linking agent only for those polymers with which it can be combined homogeneously at temperatures below 50 °C.

A potential application of UHF-sensitive peroxides is the cross-linking of thermoplastic polymers which have a small dielectric loss factor (tan $\delta \cdot \varepsilon$) (like polyethylene). Since additional blending of peroxide with polyethylene is necessary, and since a molding process during the melting operation precedes the cross-linking, only those peroxides which are still stable above 120 °C, are useful to the cross-linking process in a UHF-field; tert-butyl perbenzoate (TBPB) is one of these peroxides.

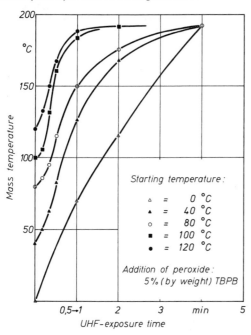

Figure 4
Rise in temperature of parts weighing 300 g which have been manufactured from PE + 5% (by weight) tert butyl perbenzoate (TBPB) as a function of UHF exposure time (2450 MHz, 2.1 kW) [7].

A mixture of TBPB blended with a nonpolar polyethylene will get hot if brought into a UHF-field (Figure 4). The maximum temperature reached depends on the concentration of the peroxide and other experimental variables. The rise in temperature of the mixture is caused by heat transfer from peroxide to polyethylene. Simultaneously with the rise in temperature of the mass, decomposition of the peroxide occurs. This decomposition cannot be described according to the laws which govern thermal decomposition of peroxides.

If only thermal decomposition occurs, then TBPB decomposes only minimally within 1.5 minutes at 100 °C. But when TBPB containing 5% PE by weight is brought into a 2.1-kW

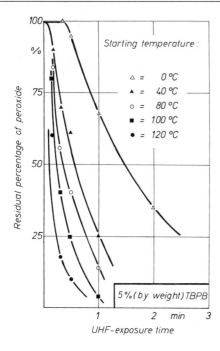

Figure 5
Decomposition of tert-butyl perbenzoate (TBPB) as a function of UHF exposure time (2450 MHz, 2.1 kW) and at various start temperatures. 300 g PE + 5% (by weight) TBPB [7].

UHF field, at a starting temperature of 0 °C, the mixture will reach 100 °C within 1.5 minutes and half of the available peroxide will decompose [Figure 5].

Two explanations are possible for this specific UHF decomposition of peroxides:

(a) Since the influence of the UHF field causes a rise in temperature of the PE-peroxide mixture, and since the heat source comprises the distributed peroxide molecules, these molecules must show a higher temperature than the average temperature of the mass indicates. Therefore, it is possible that the energy level of the peroxide molecules is sufficient to cause thermal decomposition despite the measurably lower mass temperature.

b) In an electric alternating field, a constant dipole alignment takes place in accordance with the direction of the field. This excites the peroxide molecules, and the resulting mechanical energy is transfered by collision to neighboring molecules. Mechanically over-exciting an unstable peroxide group will lead to decomposition and the generation of free radicals.

In both cases, the use of UHF energy allows decomposition of the peroxide at a lower temperature than is usually needed. Decomposition by UHF is not restricted to tert-butyl perbenzoate. At present, the sole available peroxide used in the UHF cross-linking process only is one made by Akzo Chemie. The chemical formula of this peroxide is:

Bis (tert-butyl-
peroxy) terephthalate

Bibliography to Section 6.2

[1] *N. N.:* Firmenschrift. Akzo Chemie.
[2] *Swern, D.:* »Organic Peroxides«, Bd. 1. Interscience Publishers, J. Wiley & Sons, New York-London, 1962.
[3] *N. N.:* Company Report Akzo Chemie.
[4] *Demmler, K., J. Schlag:* Kunststoffe 64 (1974), S. 78.
[5] a) *Miller, L. J., J. D. Margerum, J. B. Rust:* Journal of the SMPTE 77 (1968), S. 1177.
     b) *Oster, G., N.-L. Yang:* Chem. Rev. 68 (1968), S. 125.
[6] DEOS 26 11 349 (1976), Inst. f. Kunststoffverarbeitung an der RWTH-Aachen.
[7] *Menges, G., K. Kircher, B. Franzkoch:* Kunststoffe 70 (1980), S. 45.

6.3   Processing by Reaction Casting of Parts and Semifinished Goods from Polymethyl Methacrylate

6.3.1   General

Parts made from polymethyl methacrylate (PMMA) are produced by molding of PMMA melt, reaction casting, or processing semifinished goods. Bulky semifinished goods are usually produced by reaction casting. In 1978, 48% of the parts and semifinished goods made from PMMA in western Europe were produced by reaction casting [1].

At the moment, thermoplastic processing by extrusion is gaining importance for the manufacture of PMMA parts. The extrusion and injection molding processes are highly automated, while reaction casting is labor intensive. A particular feature of the PMMA cast polymers is their high molecular weight, approximately 5×10^6 g/mole, which gives the polymer a thermoelastic character. Injection molded polymers and extruded polymers have molecular weights of only approximately 3×10^5 g/mole. Despite the many advantages of cast polymers, reaction is limited to large sheets of material and bulky parts which could not be manufactured by any other method without great difficulty.

The manufacture of parts by reaction molding consists of polymerization of methyl methacrylate (MMA) sometimes mixed with comonomers and other additives in a mold. The exact duration of the cycle depends mostly on the geometry of the part. Thin sheets usually require a reaction time of one to several days; very thick sheets may require a carefully monitored cycle of several months. Temperature programming has to be used to control the reaction extremely carefully. Because of the great expenditure involved in reaction casting of bulky or large parts, these parts and semifinished goods are manufactured mostly by large plastic processing plants. Therefore, the number of processors is small, and manufacturers of raw material undertake the processing themselves. A summary of processing methods for parts and semifinished goods from PMMA by reaction casting is given in [2].

6.3.2   Polymerization of Methyl Methacrylate

Free radical polymerization of MMA occurs during reaction casting of bulky parts and semifinished goods; small additions of comonomer allow modifications in properties of the molding compound. Azobisisobutyronitrile(AIBN, half-life of one hour at 82 °C) or dilauroyl peroxide (half-life of 1 hour at 79 °C) are used as radical initiators.

Radical chain reaction follows the laws described in section 6.1. The following equations describe the reaction:

Chain initiation:

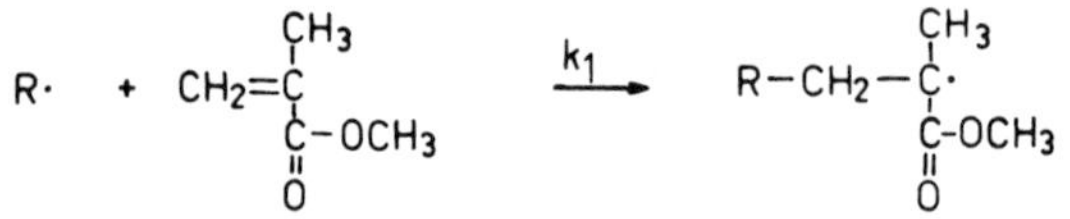

$$R\cdot\ +\ CH_2{=}\underset{\underset{O}{\overset{\|}{C}}{-}OCH_3}{\overset{CH_3}{\underset{|}{C}}}\ \xrightarrow{\ k_1\ }\ R{-}CH_2{-}\underset{\underset{O}{\overset{\|}{C}}{-}OCH_3}{\overset{CH_3}{\underset{|}{C}}}\cdot$$

Chain propagation:

$$R{-}CH_2{-}\underset{\underset{O}{C}{-}OCH_3}{\overset{CH_3}{C}}\cdot\ +\ CH_2{=}\underset{\underset{O}{C}{-}OCH_3}{\overset{CH_3}{C}}\ \xrightarrow{\ k_2\ }\ R{-}CH_2{-}\underset{\underset{O}{C}{-}OCH_3}{\overset{CH_3}{C}}{-}CH_2{-}\underset{\underset{O}{C}{-}OCH_3}{\overset{CH_3}{C}}\cdot$$

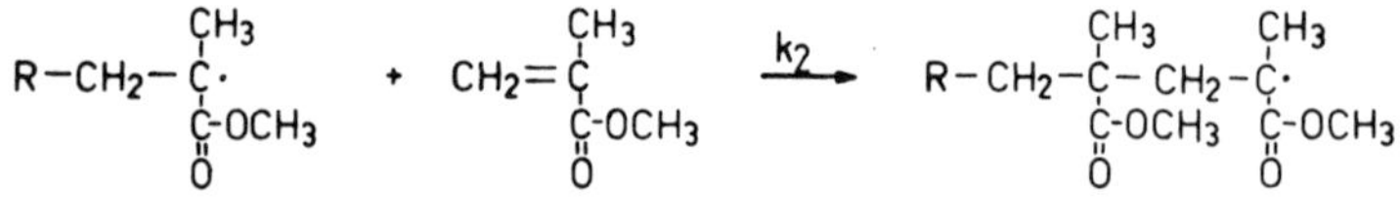

Chain termination of methyl methacrylate polymerization can occur in two ways: by combination or disproportionation. Termination by disproportionation is by far the most common [3].

Chain termination by disproportionation:

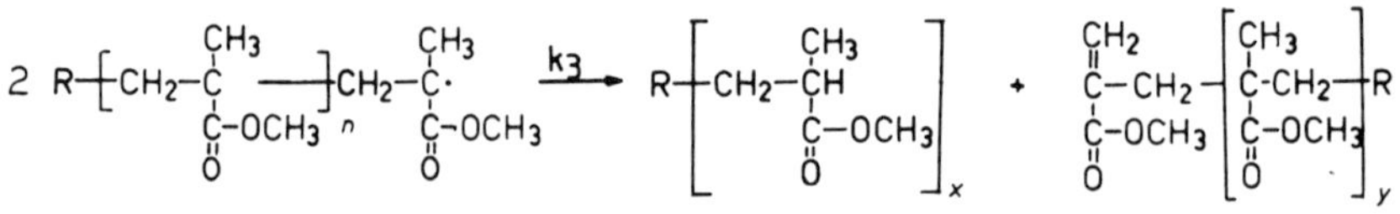

Chain termination by combination:

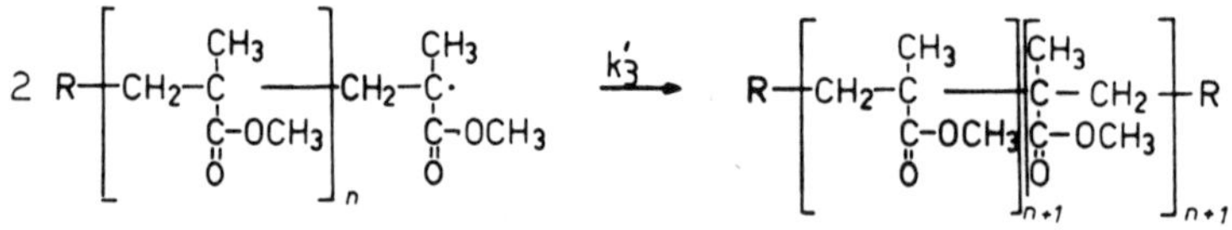

For MMA polymerization, chain termination is a bimolecular reaction. A chain termination by elimination reaction does not occur. The values for the chain propagation rate constant k_2 and chain termination rate constant k_3 and k_3' are [4]

$$\log k_2 = 6.046 - 1084\,T \qquad \text{activation energy} = 21\ kJ/mol$$
$$\log k_3 = 6.742 + 42\,T \qquad \text{activation energy} = -$$
$$\log k_3' = 6.736 - 454\,T \qquad \text{activation energy} = 8.4\ kJ/mol$$

The rate constant of the starting reaction (k_1) depends on the temperature and the type of radical generator. If the kinetic constants are known, the course of reaction can be calculated. The laws described in section 6.1 are valid only for the beginning of the polymerization reaction of MMA. After 20% of the monomer has been reacted a gel effect sets in (Trommsdorff effect) [5, 6].

The probability of two chain radicals colliding decreases with increasing chain length, since the longer chains are less mobile. At the onset of polymerization the mobility of the monomer molecules does not change, but the limited mobility of the radical structure results in the increase of the overall rate of polymerization. The gel effect is reflected in an increased rate of reaction and in higher molecular weights.

When the monomer conversion is between 40 and 80% the average life of the radical chains reaches values of over 200 times the initial one [7]. Considering the decreasing monomer concentration the rate of reaction shows its maximum at 50–70% conversion. Further conversion leads to a rapid decrease in the rate of overall reaction [Figure 6]. In the 10–50% conversion range restriction in monomer mobility is so small that it does not affect the rate of propagation. The termination constant, however, decreases noticeably. At 45% polymerization, the above-stated activation energies of the chain termination reaction (0 or 8.4 kJ/mole, resp.) increases to an apparent value of 147 kJ/mole.

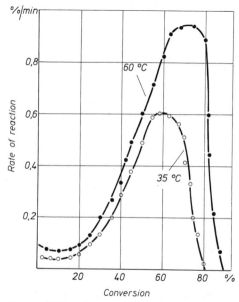

Figure 6
Rate of reaction as a function of degree of conversion [8].

Above 50% polymerization, the propagation reaction becomes increasingly slower. Molecular reactions are rate-determining at low conversions. At higher conversions, polymerization is controlled by diffusion. At very high conversions, polymerization stops despite the presence of monomer. Tempering of the parts accelerates diffusion of the monomer molecules and so allows continued polymerization. Therefore, it is necessary to heat cast PMMA parts after the reaction.

The gel effect indeed raises the rate of reaction and the kinetic chain length but difficulties may arise because of the resulting higher heat of reaction per unit time. In practice, changing the temperature of the molds will correct that problem. The gel effect begins even earlier if cross-linking comonomers are being used in the polymerization. This has to be taken in consideration when selecting the temperature of the molds.

In practice, there is usually a preliminary induction period before the actual polymerization. During this period the oxygen dissolved in the monomer reacts according to the following equation:

$$R-CH_2-\overset{\overset{\displaystyle CH_3}{|}}{\underset{\underset{\displaystyle O}{\parallel}}{\underset{|}{C}}-OCH_3}\cdot \quad + \ O_2 \quad \longrightarrow \quad R-CH_2-\overset{\overset{\displaystyle CH_3}{|}}{\underset{\underset{\displaystyle O}{\parallel}}{\underset{|}{C}}-OCH_3}-O-O\cdot$$

$$\xrightarrow{\ +\ MMA\ } \quad R-CH_2-\overset{\overset{\displaystyle CH_3}{|}}{\underset{\underset{\displaystyle O}{\parallel}}{\underset{|}{C}}-OCH_3}-O-O-CH_2-\overset{\overset{\displaystyle CH_3}{|}}{\underset{\underset{\displaystyle O}{\parallel}}{\underset{|}{C}}-OCH_3}\cdot \quad \xrightarrow{\ +\ O_2\ } \quad etc.$$

This reaction delays the chain initiation. But these polymer peroxides are able to split and act as polymerization initiators. Dissolved molecular oxygen acts as an inhibitor.

6.3.3   Manufacture of PMMA Parts and Semifinished Goods

High heat of polymerization (54 kJ/mole, which is 544 kJ/kg) as well as high volume shrinkage (23% of pure MMA) impair reaction casting of PMMA. Both effects lead to restriction of PMMA reaction casting to only special applications under specific and elaborate reaction conditions.

One can do little to reduce the high heat of reaction and shrinkage. During polymerization of MMA/PMMA the viscosity of the reaction material increases quickly (see Figure 18) and makes casting processing very difficult even at 25–30% conversion. Therefore only a very small percentage of the total heat of reaction and volume shrinkage can be eliminated in a separate preliminary process.

Reaction casting has to take the above-mentioned phenomena into consideration. Only certain molds which are able to equalize the shrinkage can be used. Polymerization in volume-variable molds is preferred. Sheets are cast in cells made of two large sheets of glass sealed by a coated rubber (e.g. PVC) gasket, making shrinkage in all directions possible. Polymerization is also possible in partially filled molds in which shrinkage occurs in only one direction. This process is used for the manufacturer of semifinished goods. Centrifugal casting allows shrinkage on the inside, which makes this process suitable for reaction casting of PMMA parts and semifinished goods. Only if shrinkage occurs uniformly on all sides will the finished part be of high quality. Both the strong dependency of the rate of conversion on the temperature and the auto-catalytic effect of the heat of reaction make uniform shrinkage possible only with precise temperature control and small uniform temperature profiles within cross sections of the PMMA part. Since heat transfer of PMMA is very low, 0.25 W/mK, the high molar heat of reaction can be dissipated only very slowly and the necessary cooling time increases in proportion to the square of the thickness of the part. This explains the long cycles.

Sometimes, because "hot spots" develop, the rate of reaction accelerates to such a degree that it leads to "runaway" polymerization. Consequently, the end product will not be glass clear but will contain bubbles. This conversion time relationship has to be given careful consideration in making provisions for heating and cooling the mold.

If the polymerization reaction is allowed to occur under adiabatic conditions, the heat of polymerization will raise the temperature of the material by 250–300 °C (average spec. heat of monomer 1.47 kJ/kgK and 2.4 kJ/kgK for polymer at 280 °C) [9]. It is interesting to compare that reaction with the reaction casting of ε-caprolactam: here polymerization under adiabatic conditions raises the temperature by 70–80 °C only and results in a rigid product which can be released from the mold without cooling. Adiabatic polymerization of MMA is impossible because the decomposition temperature of the polymer will be exceeded during the process.

6.3.4   Comonomers for Methacrylate Polymers

Various comonomers (Table 13) can be used in the synthesis of methacrylates to impart characteristic properties to the end product.

Table 13   Acrylate and methacrylate monomers

Name	Formula	MW	Kp [°C]	Heat of polymerization [kJ/mol]
Methyl-methacrylate	$CH_2 = C - C - OCH_3$, O (double bond), CH_3	100,1	100,3	55,6
Ethyl-methacrylate	$CH_2 = C - C - OC_2H_5$, O, CH_3	114,1	118,8	57,8
1,4-Butanediol-dimethacrylate	$CH_2 = C - C - O(CH_2)_4 - O - C - C = CH_2$, O, O, CH_3, CH_3	226,3	260–65	
Methacrylic acid	$CH_2 = C - C - OH$, O, CH_3	86,1	161	66,2
Methacrylic amide	$CH_2 = C - C - NH_2$, O, CH_3	85,1	215	56,1
Methyl-acrylate	$CH_2 = CH - C - OCH_3$, O	86,1	80,9	78,7
Ethyl acrylate	$CH_2 = CH - C - OC_2H_5$, O	100,2	99,4	65,3
n-Butyl acrylate	$CH_2 = CH - C - O - (CH_2)_3 - CH_3$, O	128,2	147	77,5
2-Hydroxy ethyl-acrylate	$CH_2 = CH - C - O - CH_2 - CH_2 - OH$, O	116,1	60 (1 Torr)	
Acrylic acid	$CH_2 = CH - C - OH$, O	72	141	74,5
Glycidyl acrylate	$CH_2 = CH - C - O - CH_2 - CH - CH_2$, O, O (epoxide)			

For instance, cross-linking agents such as glycol dimethacrylate, and allyl methacrylate, triallyl cyanurate are used to raise the product's heat-distortion temperature and resistance to solvents. Monomers containing long-chain alcohols result in lowered heat resistance and in internal plasticization of the material.

6.3.5   PMMA Reaction Casting Resins

PMMA reaction casting resins are MMA containing resins with a certain percentage of dissolved PMMA which gives the resin a viscosity suitable for reaction casting. Processing is the same as for UP resins. PMMA casting resins may contain different comonomers and may be mixed with additives. An exact distinction between casting resins and resins which are used to process sheets or bulky parts is not possible. Both resins are converted to a polymer by radical chain polymerizations.

Commercially available PMMA cast resins show viscosities of approximately 2500 mPa.s at 20 °C. These polymers are suitable for the manufacture of sheets, rotation parts by centrifugal casting embedments, adhesives, and laminating resins. This material is used mainly to make bathroom fixtures (bath tubs and sinks, etc.), skylights, and partitions and also for traffic signs.

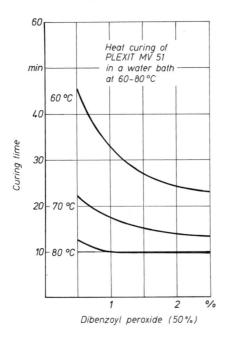

Figure 7
Heat curing of methacrylate cast resin (Plexit MV 51). Curing time as a function of temperature and content of dibenzoyl-peroxide (50%) [11].

Because of the great difference between the refractive indices for the two components, light transmission and transparency are low if common E-glass is used for reinforcement. Copolymerization with styrene and the use of a suitable grade of glass can influence the refractive index of the matrix, making possible the production of transparent parts reinforced with glass. Selecting the processing temperature and the type and concentration of the initiator allows the polymer processor to control the processing and curing time (see Figure 7).

Dibenzoyl peroxide, dilauroyl peroxide, tert.-butyl peroctoate, methyl ethyl ketone peroxide, and azobisisobutyronitrile are suitable free radical initiators for the thermosetting process. The use of an accelerator during the low temperature curing process (see Section 6.4.7) lowers the activation energy for the decomposition of peroxide to a point where an accelerated formation of free radicals is possible at room temperature [10].

Most PMMA casting resins are used to manufacture parts which are transparent. It is therefore important to use only activators that do not cause any discoloration.

Bibliography to Section 6.3

[1] *N. N.:* Kunststoffe 69 (1979), S. 496–530.

[2] *Vieweg, R., F. Esser:* »Polymethacrylate«, Carl Hanser Verlag, München, 1975.

[3] *Bevington, J. C., H. W. Melville, R. P. Taylor:* J. Polym. Sci., 12 (1954), S. 449, 14 (1954), S. 463.

[4] *Houwink, R., A. J. Staverman:* »Chemie und Technologie der Kunststoffe«, Bd. 1, 4. Auflage, Akademische Verlagsgesellschaft, Leipzig, 1962, S. 255.

[5] *Trommsdorff, E., H. Köhle, P. Lagally:* Makromol. Chem., 1 (1948), S. 169.

[6] *Schultz, G. V., G. Harborth:* Makromol. Chem., 1 (1947), S. 106.

[7] *Hayden, P., H. Melville:* J. Polym. Sci. *43* (1960), S. 201.

[8] *Burnett, G. M., G. L. Duncan:* Makromol. Chemie 51 (1961), S. 154.

[9] *Bares, V., B. Wunderlich:* J. Polym. Sci., Phys. *11* (1973), S. 861.

[10] *Paul, D. R., D. W. Fowler, J. T. Housten:* Appl. Polym. Sci. 17 (1973), S. 2771.

[11] *Röhm GmbH.:* Firmenschrift.

6.4   UP Resins

6.4.1   UP Resins as Monomers for the Manufacture of Plastics

UP resins are a mixture of unsaturated polyesters (MW 1000–2000 g/mole) and a copolymerizable, low molecular weight monomer. At room temperature these unsaturated polyesters are either highly viscous, or low-melting brittle solids, or waxy components, which are dissolved in a low molecular weight monomer. A small amount of monomer acts as a plasticizer for the unsaturated polyester. UP resins alone are only polyester intermediates, which must be distinguished from high molecular weight thermoplastic polyesters. The only thing common to both products is that the individual low molecular weight components are linked by ester groups. The structure and chain length of the individual components differ significantly. A characteristic of the unsaturated polyester is the percentage of unsaturated ethylene groups. This makes the polyester an unsaturated monomer which can be cured by free radical copolymerization with a low molecular weight monomer. Only after this curing process is completed is a usable substance obtained.

All UP resins used in technical processes are unsaturated polyesters which contain several ethylene groups per molecule; copolymerization yields a duromer which cannot be further processed by molding, extrusion, or casting. The molding process must precede the synthesis. This puts the preparation of a UP resin into the hands of the polymer processor.

Free radical polymerization of the UP resin intermediate yields a compound which for want of a better name is called "thermoset UP resin."

Usually UP resins are used for reinforced plastics, especially those containing long fibrous filler. This yields parts which show a very high degree of tensile strength and a high modulus of elasticity. Both the resin and the polymerized material serve only as adhesives if a high percentage of filler is used.

6.4.2   Composition of UP Resins

6.4.2.1   Unsaturated Polyesters

Unsaturated dicarboxylic acids are the most common source for polymerizable double bonds in the polyester. Unsaturated diols are also known, but their double bonds show lower reactivity. Despite this restriction, a large number of different structures of unsaturated polyesters are possible. A variety of polyols, unsaturated polycarboxylic acids, and saturated carboxylic acids are suitable for the synthesis of polyesters; different proportions and degrees of condensation are used to increase the multitude of structures to a significant degree.

Table 14 α,β-unsaturated dicarboxylic acids; starting material for manufacturing unsaturated Polyesters

Formula	Name
$\begin{array}{l} HC-C{\nearrow}^{O} \\ \parallel {>}O \\ HC-C{\diagdown}_{O} \end{array}$	Maleic anhydride
$\begin{array}{l} HC-COOH \\ \parallel \\ HC-COOH \end{array}$	Maleic acid
$\begin{array}{l} HOOC-CH \\ \parallel \\ HC-COOH \end{array}$	Fumaric acid
$\begin{array}{l} CH_2=C-COOH \\ \vert \\ CH_2-COOH \end{array}$	Itaconic acid
$\begin{array}{l} CH_3-C-COOH \\ \parallel \\ HC-COOH \end{array}$	Citraconic acid
$\begin{array}{l} HOOC-C-CH_3 \\ \parallel \\ HC-COOH \end{array}$	Mesaconic acid
$\begin{array}{l} Cl-C-COOH \\ \parallel \\ HC-COOH \end{array}$	Chlormaleic acid

α,β-unsaturated dicarboxylic acids are frequently used as reactive polycarboxylic acids. Table 14 lists the acids which are most commonly used. The reactivity of the resin is determined by the type of acid and the amount to be found in the polyester. In the manufacture of esters, various amounts of saturated dicarboxylic acids are used in addition to the unsaturated dicarboxylic acids to control the reactivity of the polyester and to determine the cross-link density and consequently the properties of the end product.

Saturated acids cause a decrease in the content of polymerizable chain double bonds. A short list of dicarboxylic acids which cannot be copolymerized is given in Table 15.

Table 15 Dicarboxylic acids which have no copolymerizable double bonds and are used for synthesis of unsaturated polyesters:

Formula	Name
o-phthalic acid structure (benzene ring with two adjacent COOH)	o-Phthalic acid
isophthalic acid structure (benzene ring with meta COOH groups)	Isophthalic acid
 HOOC—⟨ ⟩—COOH	Terephthalic acid
tetrahydrophthalic acid structure (cyclohexene ring with two COOH)	Tetrahydrophthalic acid
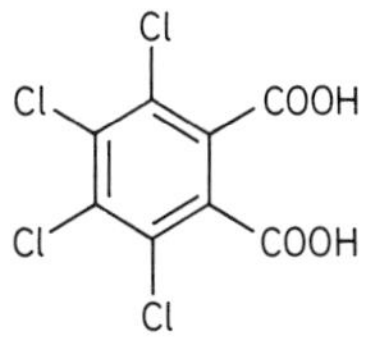 tetrachlorophthalic acid structure (benzene ring with four Cl and two COOH)	Tetrachlorophthalic acid
CH_2-COOH / CH_2-COOH	Succinic acid
$HOOC-(CH_2)_4-COOH$	Adipic acid

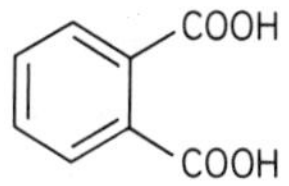

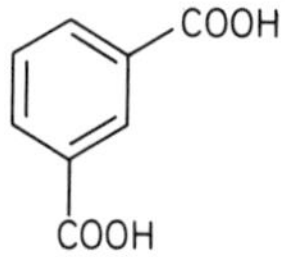

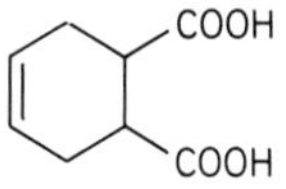

A large variety of applicable polyols are available. The same diols are used in both the preparation of polyester polyols and the PUR process.

Maleic anhydride, terephthalic acid, and ethylene glycol, for instance, are used to prepare simple polyesters. Equimolar amounts of the acids will result in a polyester with the following structure:

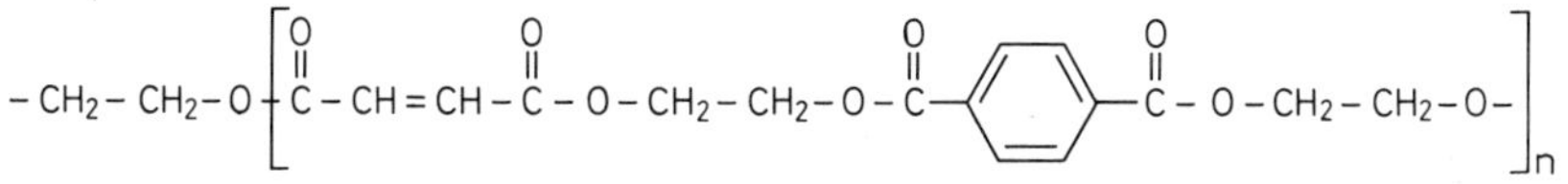

$$-CH_2-CH_2-O-\left[\overset{O}{\overset{||}{C}}-CH=CH-\overset{O}{\overset{||}{C}}-O-CH_2-CH_2-O-\overset{O}{\overset{||}{C}}-⟨\ ⟩-\overset{O}{\overset{||}{C}}-O-CH_2-CH_2-O-\right]_n$$

6.4.2.2   Copolymerizable Monomers

In principle, all monomers that under suitable commercial curing conditions are able to copolymerize with an unsaturated polyester can be used. A multitude of demands is made on the monomer. The properties of the monomer (vapor pressure, toxicity, solubility) on one hand should allow easy processing, and on the other give optimal properties to the cured end product.

Styrene, methyl methacrylate, and methylstyrene are the only compounds that are used in large quantities. Every monomer has a specific influence on the curing process of the polyester; the number of cross-linking positions and the lengths of the cross-link bridges vary. The main task of the monomer is the linkage of the polyester molecules. The monomer

Table 16   Monomers for cross-linking of unsaturated polyesters:

Name	Formula	Shrinkage Vol.-%	Heat of poly-merization kJ/mol
Diethylene glycol bis (allyl carbonate)	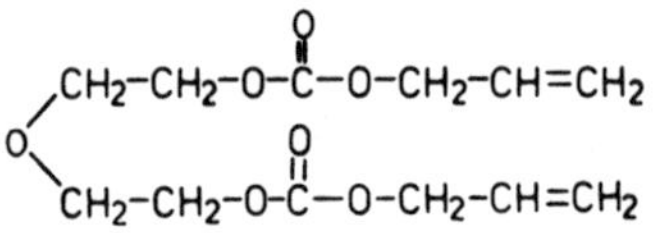 $CH_2-CH_2-O-C(=O)-O-CH_2-CH=CH_2$ / $CH_2-CH_2-O-C(=O)-O-CH_2-CH=CH_2$		
Methyl acrylate	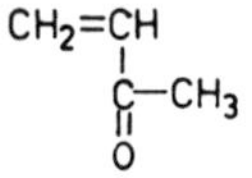 $CH_2=CH$ − $C(=O)-CH_3$		
Diallyl phthalate	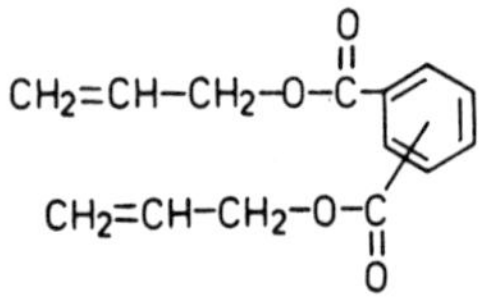 $CH_2=CH-CH_2-O-C(=O)$ / $CH_2=CH-CH_2-O-C(=O)$ (benzene ring)	11,8	
Styrene	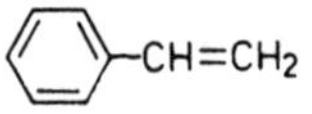 (benzene ring)$-CH=CH_2$	15–17	69,5
n-Butyl methacrylate	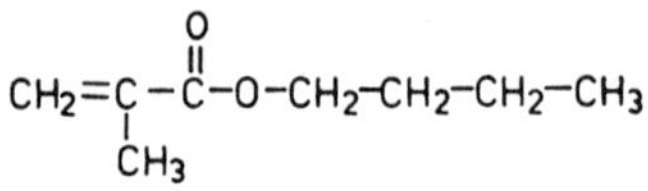 $CH_2=C(CH_3)-C(=O)-O-CH_2-CH_2-CH_2-CH_3$	14,3	56,5
Methyl methacrylate	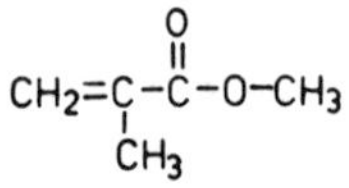 $CH_2=C(CH_3)-C(=O)-O-CH_3$	21	55,6
Acrylonitrile	$CH_2=CH-CN$	26	72,4
Vinyl acetate	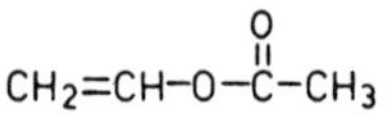 $CH_2=CH-O-C(=O)-CH_3$	27	

should homopolymerize only very little if at all. In the ideal case, an alternating copolymerization should take place, which means that in the growing polymer chain monomer bonds and chain double bonds should alternate. This would mean that even a low feed of monomers would yield maximum cross-linkage. Small shrinkage and minimal heat release are the result of low molar conversion.

The suitable monomers differ also in their ability to dissolve the polyester, in the heat of polymerization per unit weight, and particularly in the rate of polymerization.

6.4.3 Curing of UP Resins as a Radical Copolymerization

The starting materials, unsaturated polyesters and vinyl monomers, copolymerize during the curing of UP resins. The four partial reactions that occur during curing with styrene are listed in Table 17; the applicable kinetic theory for these reactions is described in Section 6.1.5.

Table 17 Intermediate steps of copolymerization between styrene and maleic/fumaric acid diesters

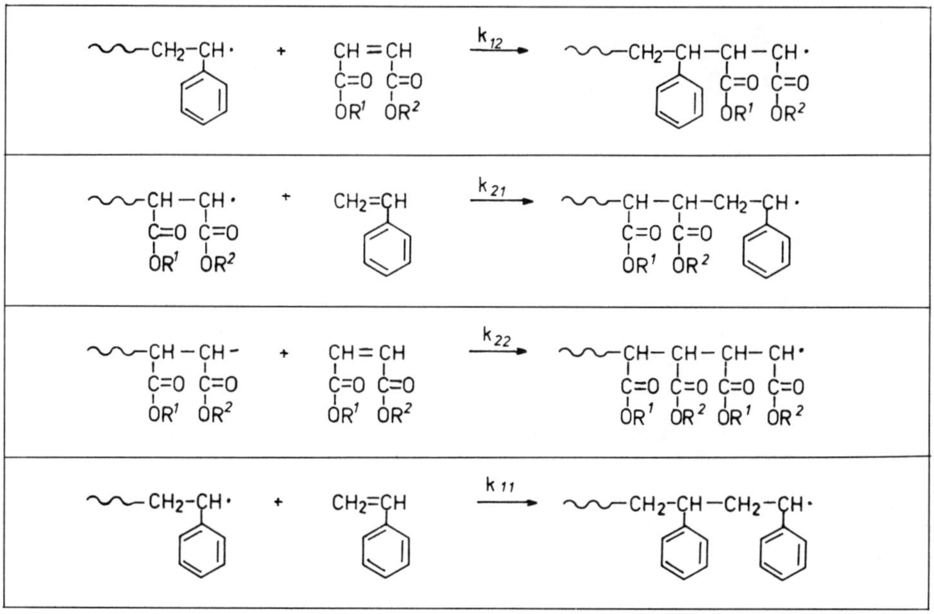

Curing of UP resins is a bulk polymerization. No other solvents are used as auxiliary compounds. In the initiation stage the monomer vinyl compound acts as a solvent; its ability to act as a solvent depends on how far the conversion has progressed.

During the reaction the viscosity of the resin changes from low viscous to rigid duroplastic. The first part of the free radical copolymerization occurs in the liquid state; toward the end of the curing process the low weight molecular styrene copolymerizes with a ethylenic unsaturated, cross-linked matrix. The kinetic theory described in chapter 6.1 is valid for the beginning of the curing reaction, but toward the end of the reaction, polymerization is controlled by the rate of diffusion. The reaction determining rate is the rate of diffusion of styrene toward the reaction site. Increasing cross-linkage limits the mobility of the chain double bonds more and more. A high degree of cross-linkage fixes the site of the chain double bonds. For steric reasons the chance of a reaction becomes less and less.

The gel effect discussed in Section 6.3.2 applies also to the curing of the UP resin. Even at low conversions the gelling of the resin lowers the mobility of the growing chains and thus increases the rate of polymerization. While chain propagation is (initially) unaffected and the probability of chain termination is reduced, the overall rate of polymerization is increased.

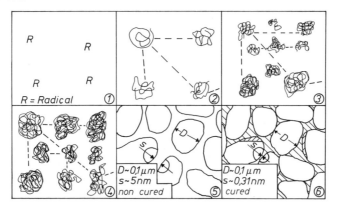

Figure 8   Change in structure of UP resins during curing (according to Demmler [1]). The originally homogenous solution becomes a two-phase system.

At the start of the reaction (Figure 8-1) statistically distributed free radicals are formed (according to Demmler [1]). These radicals start the chain reaction. First loosely formed clusters up to 0.1 µm in size appear (Figure 8-2). Finally a coarse-meshed gel is formed (Figures 8-3, and 8-4), copolymerization occurs inside the clusters, and they shrink in size; because of incompatibility of polymer and monomer, a phase separation occurs in the micro range (Figure 8-5). The addition of reactive material causes the clusters to expand to different sizes. They are held together by a (tightly) cross-linked material. The distance between the particle is approximately 0.3-1 nm (Figure 8-6).

Its insolubility makes it impossible to determine the kinetic chain length of the cured product. In order to be able to do these experiments, it is necessary to hydrolyze the bulk material. Only the polyester part of the molecule is attacked by hydrolysis. After hydrolysis of all ester groups, the structure of the molecule is as follows:

Kinetic chain length can be calculated by determining the size of the remaining hydrolytically stable molecule. Molecular weights listed in Table 18 have been determined ebullioscopically. They illustrate the kinetic chain lengths that may occur during UP resin curing.

Table 18 Molecular weight and product composition of two different polymer chains which were obtained by hydrolytic decomposition of cured UP resins [2]

Molar ratio of styrene to fumaric diester		MW of the copolymer	Units of fumaric acid per copolymer chain (statistically)
in non cured resin	in the cured resin after hydrolysis		
3,1	5,4	15000	21
0,5	1,4	11000	38

Unsaturated polyesters are polyfunctionally unsaturated monomers and produce highly cross-linked compounds; the kinetic chain length is of less importance to these cross-linked compounds than to linear molecules. But even here, very short chains may result in poor mechanical properties. Short chain lengths are formed if high concentrations of radical initiators are used.

A summary of reactions that occur during the curing of UP resins is stated in [3–9].

6.4.4 Importance of the Copolymerization Parameter for the Curing of UP Resins

The copolymerization parameter is the ratio of homopolymerization to copolymerization (monomer reactivity ratio):

Copolymerization parameter: $\quad r_1 = \dfrac{k_{11}}{k_{12}}; \; r_2 = \dfrac{k_{22}}{k_{21}}$

Table 19 lists copolymerization parameters between styrene or MMA respectively as a vinyl monomer and other monomers containing double bonds. Two types of polymerization are possible:

a) Alternating polymerization: The end group on radical chain monomer 1 prefers to react with monomer 2, and vice versa.

$\quad r_1 = r_2 < 1$

b) At the beginning of the polymerization, M is the preferred monomer for the reaction.

$\quad r_1 > 1; r_2 < 1$

Reaction of styrene with fumaric acid esters or, generally speaking, unsaturated polyesters with fumaric acid units, will result in a successful copolymerization. Only comparatively short sequences of one component are available in the chain. Using mixtures of components in similar molar amounts as feed makes it possible to guarantee a uniform polymer chain.

The vinyl monomer prefers to homopolymerize if a combination of MMA and fumaric acid esters (or a similar unsaturated polyester) is used or if MMA (or styrene) and maleic acid esters are the copolymer system. The polymer chains contain long sequences of only vinyl polymer units if equimolar amounts of components are used for polymerization.

The copolymerization parameters determine (in addition to the monomer concentration) the ratio in which the monomer appears in the composition of the copolymer (see Sec-

Table 19   Copolymerization parameters, according to [10]

M_1	M_2	Formula	r_1	r_2
Styrene	Maleic anhydride	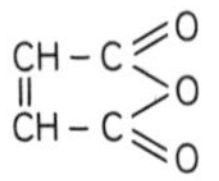	0,001	0
	Fumaric diethylester	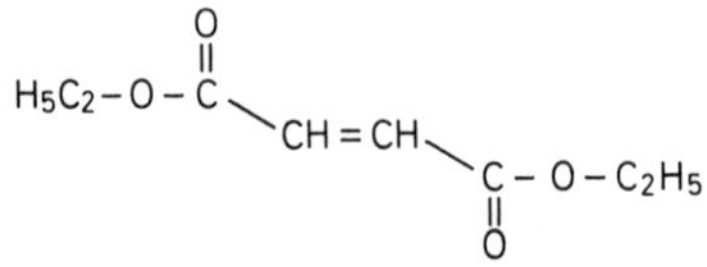	0,30	0,07
	Maleic diethylester	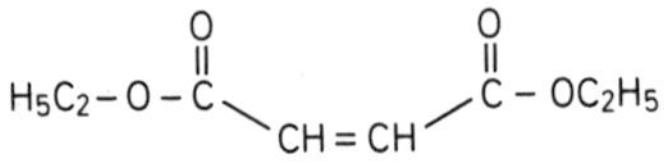	6,52	0,005
	Fumaric dimethylester	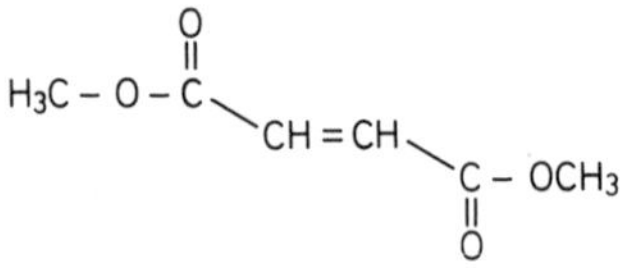	0,21	0,02
	Maleic dimethylester	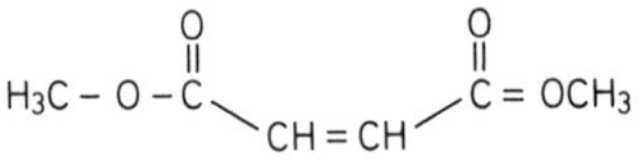	8,5	0,03
	Fumaric nitrile	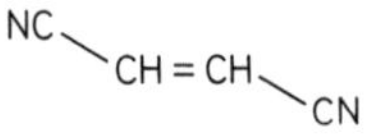	0,19	0
Methyl methacrylate	Maleic anhydride	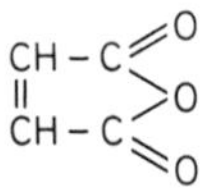	6,7	0,02
	Maleic diethylester	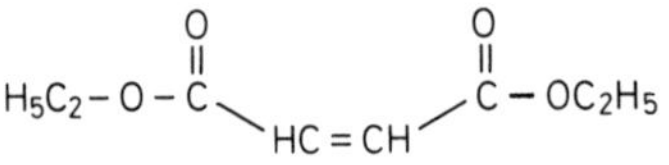	20	0
	Fumaric nitrile	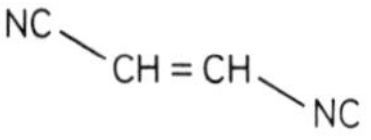	3,5	0,01
	Fumaric diethylester	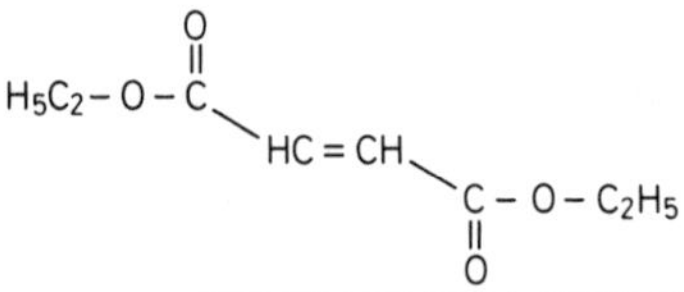	17	0,25

tion 6.1.5). The following example illustrates the difference between the vinyl monomers styrene and MMA during the curing of UP resins: What is the ratio of monomers that for both monomers will result in the same change of concentration during the polymerization process? According to Section 6.1.5, this is defined as:

$$\frac{[M_1]}{[M_2]} = \frac{r_2 - 1}{r_1 - 1}$$ $M_1 = $ Monomer 1 (Vinyl monomer)
  $M_2 = $ Monomer 2 (unsaturated polyester)

According to this equation, both monomers, styrene and fumaric acid-diethyl ester, will copolymerize in the ratio of the composition of the feed if this mixture consists of 59 mole % styrene and 41 mole % fumaric acid-diethyl ester.

The equation mentioned above cannot be used to calculate the corresponding feed composition for copolymerization of MMA and fumaric acid diethyl ester. Since $r_1 > 1$ and $r_2 < 1$, the rate of reaction for MMA is faster than that of the ester, and the result is a polymer chain with very long sequences of PMMA.

Curing of UP resins with fumaric acid esters and styrene, using approximately equimolar amounts of double bonds, results in a highly cross-linked product with short polystyrene sequences. Under the usual conditions the formation of a pure polystyrene is highly unlikely. Attempts to detect polystyrene have been unsuccessful [2].

If one uses MMA instead of styrene, the vinyl monomer reacts mostly with the radical MMA and the end product shows only a small degree of cross-linkage.

The copolymerization parameters do not allow calculation of the exact polymer composition for the curing of UP resins. The course of reaction is determined mostly by the state of gelation and the diffusion of monomer. The experimental results do not coincide with the theoretical calculations: Optimum composition of the mixture must be determined by experiment.

Table 20   Curing of a UP resin where the employed maleic acid has been isomerized to various degrees to fumaric acid; Conversion of styrene and chain double bonds as a function of the ratio of fumaric acid to maleic acid [11].

Ratio of the isomers in the resin		Conversion of styrene [%]	Conversion of the chain double bonds [%]
Fumaric acid	Maleic acid		
5,7	94,3	94,8	28,6
18,5	81,5	86,0	64,9
28,2	71,8	95,5	48,2
43,0	57,0	93,6	59,2
58,0	42,0	92,0	68,8
76,2	23,8	90,0	71,4
100,0	0	86,1	80,2

Maleic acid diesters are only slightly reactive. Table 20 lists analyses of several cured resins in which fumaric acid is replaced gradually by maleic acid diester. Under the given conditions of reaction, a resin with only fumaric acid units shows polymerization of 80.2% of chain double bonds; while in a resin with 94.3% maleic acid and 5.7% fumaric acid structure only 28.6% of the chain double bonds reacted. Almost total reaction of styrene occurs in all cases.

Only in combination with fumaric acid units will polymerization of styrene result in highly cross-linked products. Styrene-cured resins are wide-meshed cross-linked if the chain double bonds belong mostly to maleic acid diesters. (Similar concentration of chain double bonds is taken into consideration). Copolymerization of styrene and maleic acid diesters results mostly in homopolymerization of styrene and long chains that are not cross-linked.

In the ternary system of styrene, maleic acid, and fumaric acid, nine independent rate constants and nine reactions have to be considered. Since radical maleic acid and fumaric acid units do not react with each other or with themselves, one can omit the three reaction probabilities.

A mixture of styrene and MMA increases the amount of MMA in the cross-linked polyester molecule. Styrene acts as a "copolymerization agent" because of the very favorable copolymerization parameter for MMA and styrene.

$$\left.\begin{array}{l} \text{Monomer 1-MMA} \\ \text{Monomer 2-styrene} \end{array}\right\} \quad r_1 = 0.46;\ r_2 = 0.52$$

If the first step in a polymerization consists of addition of a chain double bond to a free radical unit, then it is preferable to carry it out with a styrene radical.

Variations in the molar ratio of styrene to chain double bonds will result in an altered composition of the polymer (see Table 21). Neither styrene nor all the chain double bonds show complete conversion. Approximately a 1:1 molar ratio of styrene to chain double bonds yields optimum conversion of styrene during the curing of the resin mentioned below (Table 21); but only 85% of the fumaric acid double bonds react. If the molar ratio of styrene to chain double bonds is 2:1 then 95% of both kinds of double bonds will react. Similar conditions exist for other UP resins with fumaric acid components. This explains why the most widely used molar ratio for both kinds of double bonds is 2:1 in most resins.

Table 21   Curing of an unsaturated polyester made from fumaric acid, adipic acid, and hexane diol (acid number 26.8) containing double bonds of 0.264 mol per 100 g polyester with various amounts of styrene [12]

Number of moles of styrene prior to the reaction	Conversion of styrene in mol-% in relation to the charge amount	Conversion of fumaric acid in mol-% in relation to the charge amount
0,289	85,80	38,13
0,393	93,34	57,80
0,478	93,36	74,54
0,549	97,00	84,22
0,611	95,40	94,61
0,710	93,00	93,83
0,786	91,81	97,77
0,872	91,81	94,42
0,917	90,84	99,22
0,936	91,65	99,33

6.4.5   Curing of UP Resins

Conditions for curing reactions of UP resins vary greatly:

Temperature: Room temperature to 170 °C
Pressure: Ambient to about 200 bar
Environment: Contact with air, curing in molds
Radical initiation: Thermal and photolytic decomposition of radical generators

UP resins are a mixture of reactive components. In principle they are able to react at all temperatures. Initiator radicals are necessary in order to start and continue the curing process. The polymer processor is responsible for the exact timing and concentration of the

radical initiators. Premature formation of the radical initiators must be avoided. It is important to have a sufficient concentration of radicals available at all times.

A specific peroxide is useful as a polymerization initiator for only a very narrow range
of temperature (see Section 6.2.1). This explains the variety of commercially available radical initiators.

Curing reactions can be classified as:

- UV-initiated reactions
- Reactions using thermal or catalytic radical generation

Curing reactions that use radical initiators formed thermally or catalytically can be
subdivided according to the temperature necessary to start the reaction. Two extreme cases
exist: hot curing and cold curing. During the hot curing process, thermal decomposition of
radical initiators starts the formation of free radicals. In the cold curing process, the radical
formation is achieved at lower temperatures by adding compounds (accelerators) which
lower the decomposition temperature of the radical source.

6.4.6   Hot Curing of UP Resins

Before the curing process starts, a radical initiator is added to the resin. The radical initiator has to be sufficiently stable at room temperature to avoid premature formation of
radicals. At room temperature the rate of decomposition of peroxides is not zero; its value
can be calculated (Section 6.2.1). This results in a short shelf life since there is a certain slow
radical initiation even at room temperature; a simultaneous addition of inhibitors will
lengthen the shelf life. The use of inhibitors is also important during the molding of parts
which have complicated shapes which require the melt to flow a long distance, since premature decomposition of the peroxide occurs in the heated molds which may lead to gelling
before the molding process is completed. The peroxide used for initiating the process has to
be suitable for the temperatures of the mold and fast initiation of radicals is necessary. If the
decomposition temperature of the peroxide is too low, polymerization will start too early
and end too quickly. This leads to a shorter cycle, but too often the end products are parts
which still contain a high percentage of monomer.

If a radical initiator has an optimum temperature of decomposition that is higher than
the temperature of the mold, the reaction will start too late or not at all.

Shortly after the mold is filled, the resin is heated only by the heat of the walls of the
mold and no substantial decomposition of peroxide occurs. At a certain temperature (which
depends on the type of radical initiator), sufficiently fast decomposition of the radical initiator initiates an exothermic polymerization. The heat of the exothermic reaction is much
greater than the heat which was absorbed from the mold. The autocatalytic effect of the
heat of reaction suddenly increases the bulk temperature; this temperature decreases after
conversion of the reactive component.

Decomposition of the peroxide does not occur below 80 °C even when the temperature
of the mold is 140 °C and the relatively thermally unstable peroxide tert-butyl peroctoate is
used. Above 80 °C polymerization of the resin starts quickly; when the bulk temperature
reaches 150 °C the half-life of the peroxide is only 12 seconds and the peroxide decomposes
spontaneously (explosively).

At even higher temperatures, hardly any peroxide is left to initiate new radical chains. Polymerization occurs at this point only because a few free radicals are still left or because radicals that have been initiated by other methods are available. Loss of radicals leads to premature termination of the polymerization reaction. A high percentage of monomer is left, and the quality of the polymer is poor. The use of inhibitors does not change the outcome of the reaction. Inhibitors, used in the correct amount, lengthen the reaction only from the starting phase to the beginning of gelation.

The use of a thermally stable peroxide such as di-tert-butyl peroxide lengthens the curing time. The rise in the temperature vs. time curve is shifted toward higher time units (Figure 9). Measurable decomposition of the peroxide starts at 100 °C but the stability of the peroxide is high enough to initiate free radicals up to 180 °C. Only above this maximum temperature does the peroxide cease to be a source free radicals. Di-tert-butyl peroxide is able to initiate free radicals up to the completion of the curing process. Termination of reaction occurs at a high percentage of conversion, and the polymer quality is not diminished by a high residual amount of styrene.

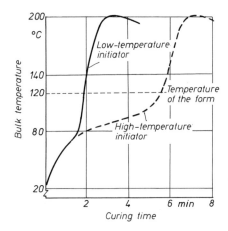

Figure 9
Schematic plot of the bulk temperature during high temperature curing of UP resins.

In the selection of a radical initiator, one has to weigh its thermal stability against the temperature range that exists in the mold. A good adjustment to temperature can be obtained by using mixtures of peroxides. A combination of high and low-temperature radical initiators is useful for fast initiation of free radicals at lower temperatures. The initiator which is stable at high temperatures allows the polymerization to continue to the end and only a small amount of unused monomer remains. In the ideal case, free radicals are initiated at a constant rate during every phase of the curing process. Chain termination by reaction of free radicals with primary radicals occurs at very high radical concentration; this leads to very short kinetic chains and an unfinished curing process.

In order to maintain a constant rate of formation of free radicals while using a single peroxide, the temperature must be constant in the resin. This can be accomplished only during the processing of very thin parts or at a very slow rate of reaction.

6.4.7   Room Temperature Curing of UP Resins

Radical initiated curing reactions that start at low temperatures, especially room temperature are used for a variety of applications. If these reactions are to be started only by radical initiators, then the use of radical initiators which have a high rate of decomposition at room temperature is necessary. These compounds have to be stored at very low temperatures; if they reach room temperature for even a very short time (for instance during shipping) they will show loss of efficiency. In addition to that, pure peroxides present a great danger if heated to the point of decomposition. Therefore radical initiators with optimum temperatures of decomposition of 20–50 °C are not very useful in the polymer processing plant.

Room temperature curing of UP resins uses peroxides to which accelerators or amines have been added to speed up the decomposition at these temperatures. This calls for the use of peroxides which are sufficiently stable at room temperature and can easily be shipped, stored, and handled. Section 6.2.1 describes their scope and reactions.

UP resins mixed with peroxides or with accelerators can be stored separately for a reasonable time. The addition of inhibitors increases their shelf life. If a peroxide-containing resin is mixed with an accelerator in styrene or UP resin, then the decomposition temperature of peroxide is lowered and room temperature is sufficient to start decomposition. Consequently, the UP resin will start to set at room temperature.

It is very important to keep peroxide and accelerator separated. Mixing of the pure components is followed by spontaneous decomposition of the peroxide. Generally, the accelerator is dissolved in styrene and then mixed with the resin. The accelerator solution is the last component to be added, shortly before the process starts. As soon as the accelerator is added initiation of free radicals starts. Inhibitors must be used to avoid premature gelation of the resin. The inhibitors trap the available free radicals until the addition of fillers or reinforcing fibers and the molding process are completed.

Generally, room-temperature curing of UP resins consists of processes which start polymerization at ambient temperature. The temperature at which the use of accelerators is no longer necessary is the dividing line between room-temperature and hot curing. Accordingly, a process with reaction temperatures of approximately 50 °C would still be called room-temperature curing.

Cobalt and vanadium salts which are soluble in the resin and amines are effective accelerators. Table 22 shows a list of these compounds. Metal compounds and amines show a different degree of activity. Metal compounds are true catalysts which work on the principle of redox reactions. Amines are reaction components (see Section 6.2.1). Metal salts are used in catalytic amounts only. Most widely used is a solution containing 1% catalyst in styrene. Only 0.1–3% of this solution is added to the resin; this is a molar ratio of 1:0.01 of peroxide to accelerator. Amines are not true catalysts; they act as accelerators by reacting with the peroxide; this reaction produces free radicals, and larger amounts of amines are therefore necessary. The ratio of peroxide to accelerator is 1:0.1–1.0, preferably 1:0.5. Since larger amounts are added, pure components are usually used.

Peroxide and accelerator have to be carefully matched with respect to quality and quantity. Cobalt salts accelerate the decomposition of ketone peroxides. Vanadium salts are useful for the decomposition of peresters, perketals, and hydroperoxides. Decomposition of diaryl peroxides, whose most important representative is benzoyl peroxide, for room-

Table 22   Accelerators for room-temperature curing of UP resins

Abbreviation	Name	Formula
DMA	N,N-Dimethylaniline	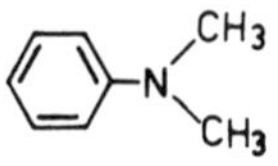 (phenyl)–N with CH_3, CH_3
DMoT	N,N-Dimethyl-o-toluidine	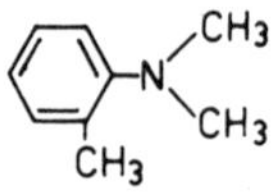 (phenyl with CH_3)–N with CH_3, CH_3
DMpT	N,N-Dimethyl-p-toluidine	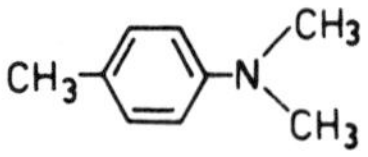 CH_3–(phenyl)–N with CH_3, CH_3
DEA	N,N-Diethylaniline	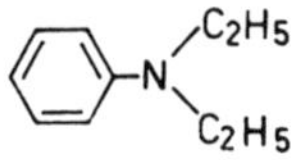 (phenyl)–N with C_2H_5, C_2H_5
EBA	N-Ethyl-N-benzylaniline	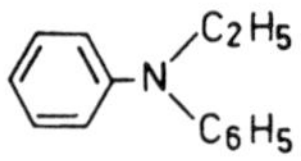 (phenyl)–N with C_2H_5, C_6H_5
DBA	Dibenzylaniline	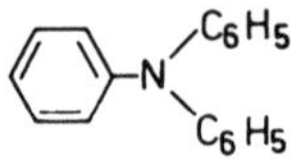 (phenyl)–N with C_6H_5, C_6H_5
DiPA	N,N-Diisopropanolaniline	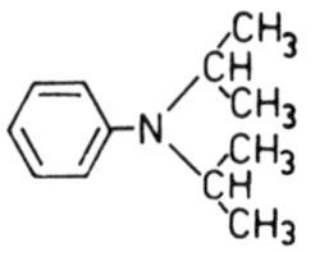 (phenyl)–N with CH_3–CH–CH_3, CH_3–CH–CH_3
Co-octoate	Cobalt octoate	$(C_7H_{15}-COO^-)_2\ Co^{++}$

temperature curing is best accelerated by the addition of amines. Changing the ratio of per-oxide to accelerator will allow variation in the rate of radical initiation. There is an upper limit to the useful amount of accelerator, because increasing amine concentrations decrease the amount of active free radicals.

Reactions which are accelerated by cobalt salts show nearly the theoretical value of 50% of polymerization-initiated radicals. This value drops to 10–20% for reactions that are accelerated by amines [13]. Processing recipes have to take into consideration the different utilization of the peroxide. Since the reaction between amines and peroxide produces com-pounds with radical character in which the accelerator is incorporated; the accelerator will also be incorporated into the polymer chain.

Even when the amine is incorporated into the polymer chain, it still has a small effect on the decomposition of the peroxide [14]. The accelerating effect of the amine decreases slowly. Radicals formed by amine acceleration are highly active and react immediately with polymerizable components. Therefore, the concentration of these radicals is very low in the resin; only during high peroxide concentrations is it possible to detect primary radicals analytically (by electron spin resonance) [15].

6.4.8 Ultraviolet Curing with Photoinitiators

In order to use UV light for the curing of UP resins, a special radical initiator which will decompose photolytically has to be added (Section 6.2.2). After forming, the UP resin is exposed to a UV light source. As with the other methods, the curing rate depends on the amount of free radicals initiated per unit time. The light intensity decreases logarithmically with increasing depth of infiltration; the rate of curing decreases accordingly.

Commercially available UV lights with maximum wavelengths of 366 nm are unable to penetrate anything but very thin layers of polyester resins (see [16]). For quite a while UV curing was used only for the setting of thin surface layers. Pigmented polyester varnishes cannot be cured totally because UV rays are able to penetrate only transparent layers. It was also believed that UV rays could not be used on UP resins reinforce with glass fibers since they are not transparent enough, but more recent studies have shown that a certain amount of glass fiber is acceptable [17]. Even parts with a thickness of 4–10 mm can be cured with UV light with acceptable short curing times if sufficient active initiators are present [18].

Mixtures of UP resins and photoinitiators can be stored easily in opaque containers. No additional component is necessary for the processing. Parts that have been cured by UV light show a lower amount of residual styrene than parts cured thermally using peroxides. Any UV stabilizer present, however will delay the curing process; this presents a definite disadvantage.

The kinetics for curing by UV light are the same as for the thermosetting process that uses peroxides, but the rate of free radical initiation is different. For the photosetting process, the rate of free radical initiation V_R is proportional to the quantum yield H of the initiator and to the amount of absorbed light I_a [19] (see section 6.2.2).

$$v_R \approx \text{H} \cdot I_a \qquad I_a = I_o \left(1 - e^{\varepsilon \cdot a \cdot [c]}\right)$$

$$v_R \approx \text{H} \cdot I_o \left(1 - e^{\varepsilon \cdot a \cdot [c]}\right)$$

I_a = light intensity at depth a
I_o = light intensity at the surface
$[c]$ = initiator concentration
a = depth
ε = extinction coefficient of the initiator
ⓗ = quantum yield
V_R = rate of radical formation

The equation for the initiating rate V_1 of the photopolymerization is as follows:

$$v_1 \approx \text{H} \cdot I_o \left(1 - e^{\varepsilon \cdot a \cdot [c]}\right) \cdot [M]$$

V_1 = rate of initiating reaction
$[M]$ = concentration of monomer

Light absorption makes the rate of the curing process dependent on the location inside the part; the result is an inhomogeneous curing. Of course this problem exists only for thick objects.

6.4.9   Inhibitors as Processing Aids

Inhibitors are used to render inactive all free radicals which are initiated either during storage or at certain steps during processing when polymerization is not yet desirable.

Inhibitors are processing aids which do not react with the polymer; their activity occurs prior to the actual processing reaction of the polymer. Inhibitors do not belong to a single group of chemical compounds. Fillers, water, and oxygen in the air can act as inhibitors. Table 23 shows a summary of inhibitors.

Table 23   Inhibitors

Name	Formula	Name	Formula
Hydroquinone	 HO—⟨⟩—OH	3,5-di-tert.-Butyl-catechol	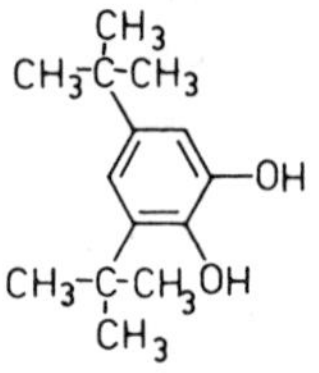 (structure)
p-Benzoquinone	 O=⟨⟩=O		
Catechol	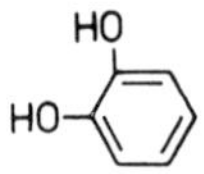 (structure)	2,5-di-tert.-Butyl-hydrochinone	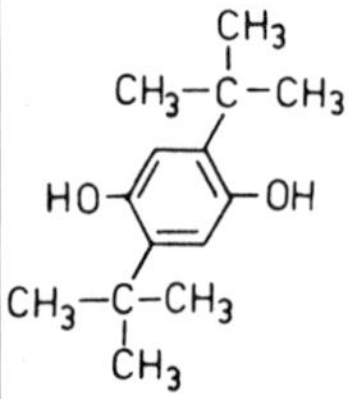 (structure)
tert. Butyl hydroquinone	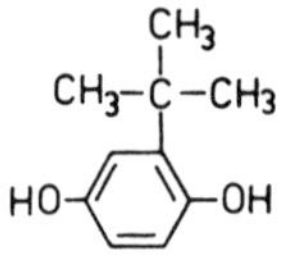 (structure)	Chloranil	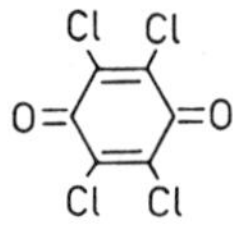 (structure)
4 tert. Butyl catechol	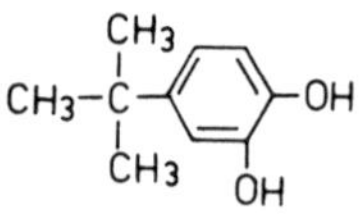 (structure)	Hydroquinone mono-methylether	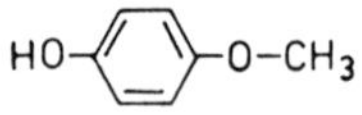 HO—⟨⟩—O–CH₃

Inhibitors are not used only for the curing of UP resins. Generally, they are used whenever free-radical-induced reactions may occur prematurely. Therefore, most of the polymerizable monomers are stabilized by inhibitors when they are delivered to the processing plant. Sometimes inhibitors are used in polymers that can be cross-linked be free radicals.

The inhibiting effect of additives, such as fillers, is difficult to describe quantitatively, since the effective compounds are not quite defined and many overlapping effects exist. In addition to that, several inhibiting reactions are surface reactions, and surface textures vary greatly. This makes it impossible to describe the effect theoretically. It cannot be forgotten that commercial resins contain additives which may act as inhibitors and their effect must be determined experimentally.

The effect of an inhibitor is demonstrated using hydroquinone as an example [20]:

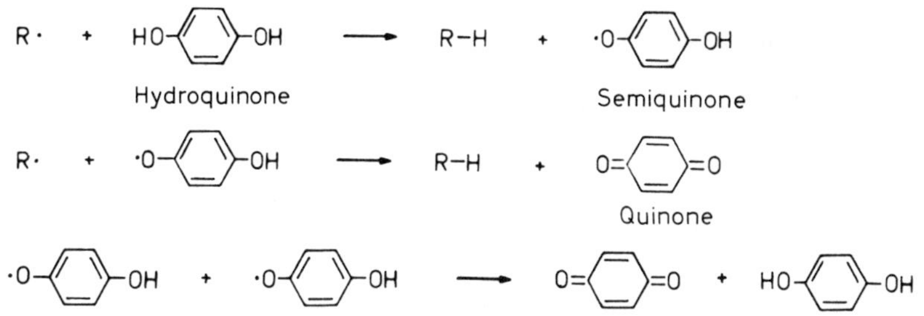

R· + HO—⟨ ⟩—OH ⟶ R–H + ·O—⟨ ⟩—OH

Hydroquinone   Semiquinone

R· + ·O—⟨ ⟩—OH ⟶ R–H + O=⟨ ⟩=O

Quinone

·O—⟨ ⟩—OH + ·O—⟨ ⟩—OH ⟶ O=⟨ ⟩=O + HO—⟨ ⟩—OH

(Disproportionation reaction)

Prematurely formed free radicals react with hydroquinone, yielding quinone. Quinone itself will react with free radicals according to the following equations:

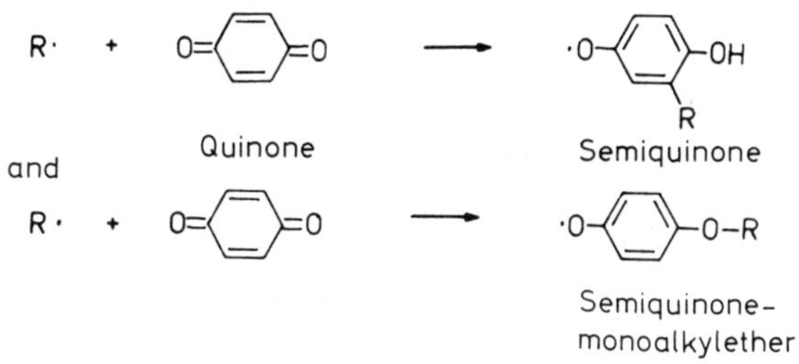

R· + O=⟨ ⟩=O ⟶ ·O—⟨ ⟩—OH (R)

     Quinone   Semiquinone

and

R· + O=⟨ ⟩=O ⟶ ·O—⟨ ⟩—O–R

Semiquinone–monoalkylether

The newly formed semiquinone continues to react:

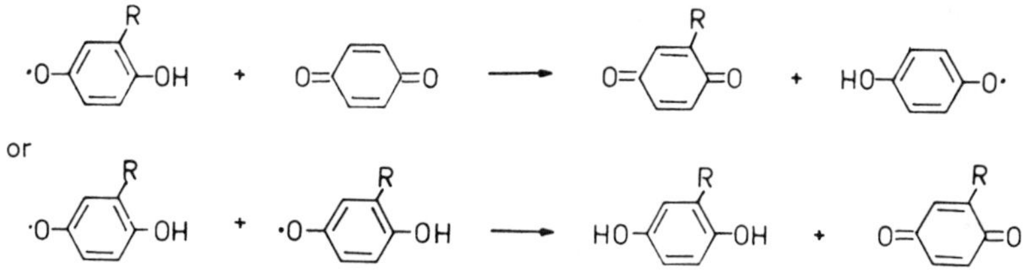

·O—⟨ ⟩(R)—OH + O=⟨ ⟩=O ⟶ O=⟨ ⟩(R)=O + HO—⟨ ⟩—O·

or

·O—⟨ ⟩(R)—OH + ·O—⟨ ⟩(R)—OH ⟶ HO—⟨ ⟩(R)—OH + O=⟨ ⟩(R)=O

Semiquinone ether is not reactive and converts easily to hydroquinone diether. Other inhibitors with quinone structure react similarly. Phenols react by losing the phenolic hydrogen atom. Amines react by forming ammonia compounds. Various inhibitors differ in the way in which reactions occur and in the rate of these reactions. It is possible to increase the rate of reaction by adding coaccelerators (i.e., hexamethylphosphoric acid triamide) [20].

The "ideal inhibitor" should react in the following manner: The reactivity of the inhibitor is so high that it reacts immediately with every prematurely formed free radical, thus making it impossible for a free radical to initiate the polymerization. Polymerization will start only after total consumption of the inhibitor.

In reality, the reaction of the inhibitor is in competition merely with the initiation of the chain reaction:

$$R \cdot \Big\langle \begin{array}{l} \nearrow + \text{Monomer} \xrightarrow{k_1} \text{second reactive radical} \\ \\ \searrow + \text{Inhibitor} \xrightarrow{k_2} \text{chain termination} \end{array}$$

(1)

(2)

k_1 and k_2 are rate constants.

Reaction of the primary radicals is a second degree-reaction. The rate of the radical-consuming reaction is proportional to the radical concentration multiplied by the concentration of monomer and initiator:

$$-\frac{d\,[R\cdot]}{dt} = k_1 \cdot [R\cdot] \cdot [\text{Styrene}] \tag{1a}$$

$$-\frac{d\,[R\cdot]}{dt} = k_2 \cdot [R\cdot] \cdot [\text{Inhibitor}] \tag{2a}$$

The following quotient represents the ratio which determines both reactions:

$$\frac{\text{chain start}}{\text{Inhibition}} = \frac{k_1 \cdot [\text{Styrene}]}{k_2 \cdot [\text{Inhibitor}]}$$

Since the concentration of inhibitor is small in comparison with the concentration of styrene, the value of k_1 must be much higher than k_2. Gelation of the resin does not occur after the first or even the second chain addition. An inhibitor is able to react with every chain radical regardless of the length of the chain. This makes the probability very high that a chain propagation will be terminated after only a few steps.

Since the inhibitor is active during the polymerization, it will affect the quality of the finished product. Slowly reacting inhibitors produce short kinetic chains and polymers that are not completely cross-linked. One must take into consideration all the factors mentioned above (the resin, reaction conditions and properties of the radical initiator) when selecting suitable inhibitors.

6.4.10   Reaction Time Performance during the Curing of UP Resins

The conversion of polymerizable components during adiabatic reactions can be measured by the graph of temperature vs. reaction time. Another method is the experimental analysis of unconverted reactive groups during the reaction. Gas chromatography is used for determining the amount of styrene; the number of chain double bonds is determined fairly accurately by infrared spectroscopy. Figure 10 illustrates a typical conversion vs time graph; it shows the conversion of chain double bonds during a cold-setting process.

Factors which influence the rate of reaction are:

- Type of active chain structure
- Type of monomer
- Temporary concentration of free radicals
- Bulk temperature

The rate of initiation of free radicals is directly proportional to the temperature. The rate of polymerization is also affected by the bulk temperature. High temperatures also accelerate the curing process, since the radical concentration increases with temperature; this

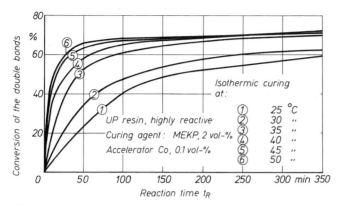

Figure 10   Reaction of chain double bonds of UP resins during curing. MEKP equals methyl ethyl ketone peroxide (curing agent); catalyst: 0.1% vol. of Co salt solution (1%) [21].

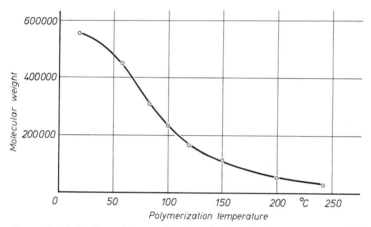

Figure 11   Molecular weight as a function of polymerization temperature during polymerization of styrene [22].

leads to not only a larger number of propagating chains, but also an increased rate of monomer addition. Also, high temperatures cause the formation of short kinetic chains. Chain propagation and chain termination are influenced differently by the temperature. Higher temperatures result in earlier chain termination and therefore lower molecular weights. This is illustrated in an example of polymerization of styrene in Figure 11.

Using large amounts of radical initiators will result in additional chain termination by primary free radicals and even shorter chains. If the concentration of initiators is too high, the parts will be incompletely cured and of poor quality.

6.4.11   Shrinkage and Heat of Reaction

Type and concentration of the polymerizable components determine the degree of shrinkage during the curing of UP resins (see Figures 12 and 13). Shrinkage is directly proportional to the amount of converted polymerizable components: Each type of double bond adds a certain amount of shrinkage to the total shrinkage.

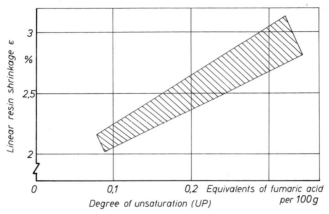

Figure 12   Increase in volume contraction of the pure resin as a function of increasing number of double bonds [21].

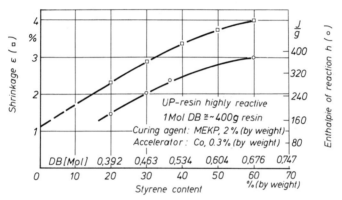

Figure 13   Increase in linear contraction and enthalpy of reaction as a function of percentage (by weight) of styrene (2% of methyl ethyl ketone peroxide and 0.3% by weight of Co salt solution) [21].

Table 16 lists the percentage of shrinkage for polymerization of pure vinyl monomers. Direct determination of the shrinkage of pure, unsaturated polyester is not possible, since pure maleic acid and fumaric acid diester do not polymerize to any noticeable degree. Indirect determination is possible: Total shrinkage of resins containing the same polyester molecule (but different monomer quantity) is plotted against percentage of monomer and then extrapolated to 0% monomer [21] (see Figure 13).

The polymer processor does not have much control over the problem of shrinkage. If he tries to decrease the amount of shrinkage by reducing the total content of double bonds in the resin, the mechanical properties of the product will change. One method that is used frequently is to add larger amounts of filler. This does not change the amount of shrinkage in relation to the amount of resin, but the absolute amount of shrinkage is reduced since the object does not contain only resin. Glass fiber can also be used as a filler.

Another method to reduce shrinkage employs special additives. Swelling of these compounds allows them to absorb part of the vinyl monomer (for instance, polystyrene or PMMA) [23]. During the curing process the monomer diffuses from the additive back into the resin and takes part in the polymerization. The decreasing amount of monomer in the swollen additive particles will result in a reduction in particle volume; if the resin is not sufficiently plastic to allow this volume reduction, a void is formed around the particle. The shrinkage, which usually changes the outer dimension of the part, occurs in "low-profile" systems internally. Instead of a smaller part being produced, internal cavities are formed. In principal, the amount of total shrinkage stays the same.

Curing of a UP resin is an exothermic reaction. The heat of reaction depends on the type and concentration of polymerizable double bonds. It is not possible to change the enthalpy of reaction without changing the resin composition. The heat of reaction depends on the resin type and usually it is between 50 and 70 kJ/mole. This is enough heat to induce a strong temperature increase in the resin under adiabatic conditions. In practice, part of this heat is dissipated to the surroundings, especially if the wall thickness is small and the reaction is slow. The relatively large amount of filler which is usually present also absorbs part of the heat. Therefore the temperature increase is rarely enough to cause a real problem.

The curing process of the resin is substantially slower if α-methyl-styrene is substituted for a certain percentage of styrene; this allows better heat dissipation to the filler and the surroundings. In the polymer, α-methylstyrene and styrene alternate regularly along the chain. Pure α-methylstyrene copolymerizes very slowly with unsaturated polyesters. Addition of small amounts of α-methylstyrene will result in polymers with a higher degree of hardness, better tensile strength and elastic modulus, but less toughness than polymers containing pure styrene.

6.4.12   The Effect of Air on the Curing Process

Air can get inside the UP resin through mechanical entrapment or by air contact with the surface of the resin. Pressure is used to remove the air quickly during the curing process in molds. During most laminating processes, when the curing occurs in air, oxygen reacts with the surface of the curing resin. The oxygen present in air can have a biradical configuration:

$\cdot O - O \cdot$ Oxygen in the biradical form

and this form of oxygen is able to react with the radicals of the molding compounds during curing. Primary radicals react with oxygen [24]. These newly formed radicals react with styrene very slowly. A reaction of styrene radicals with oxygen and consequent copolymerization are also possible [25]. This reaction leads to polymers with peroxide linkages which decompose at elevated temperatures and form new radicals:

In practice, oxygen inhibition results in incomplete curing or setting of the surface of the molded part. This undesirable effect can be avoided by using special types of resin that allow fast curing in the presence of air. These resins are used as surface coating (top layer). They are manufactured by using tetrahydrophthalic acid, glycerin, and tris (2-carboxylethyl) isocyanurate, and obviously their activity depends on the increased cross-linking possibilities.

Surface coating with a protective layer is another method for keeping oxygen out. If small amounts of paraffin wax are added to the resin, the paraffin will rise quickly to the surface since it is incompatible with the matrix. The surface of the molded part is covered with a thin layer of paraffin which prevents contact with air.

6.4.13   Reactions with the Filler

UP resins are usually processed using fillers; glass fibers are the most important fillers. The interface between resin matrix and filler is usually a region of weakness.

In order to improve the mechanical properties of the cured resin, one uses a coupling agent between the resin and filler material [26]. Typical coupling agents are silanes; their silanol groups enable them to show an affinity to the filler, and the organic group allows compatibility with the matrix. The boundary layer matrix/filler is replaced by the composite matrix/coupling agent/filler. This shows a much higher degree of adhesion between filler and matrix than in any other single system.

Coupling agents that are used in connection with UP resins are based on the following compounds:

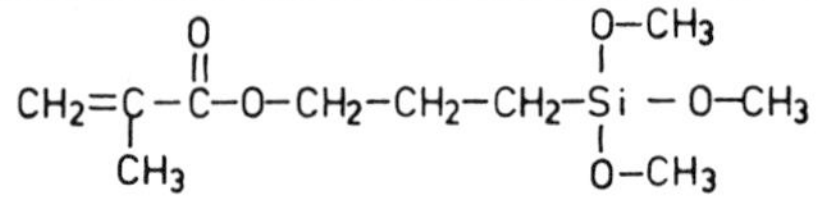 $CH_2{=}C{-}C{-}O{-}CH_2{-}CH_2{-}CH_2{-}Si{-}O{-}CH_3$ with O (double bond), CH_3, and $O{-}CH_3$, $O{-}CH_3$ substituents	γ-Dimethylacryloxipropyl-trimethoxysilane
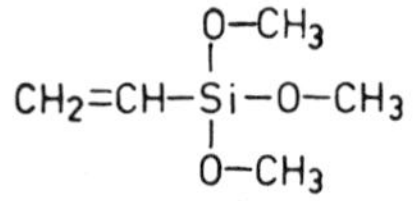 $CH_2{=}CH{-}Si{-}O{-}CH_3$ with $O{-}CH_3$ and $O{-}CH_3$ substituents	Vinyl-trimethoxysilane
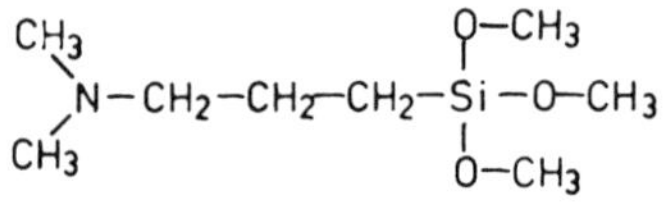 CH_3, CH_3 $\,N{-}CH_2{-}CH_2{-}CH_2{-}Si{-}O{-}CH_3$ with $O{-}CH_3$ and $O{-}CH_3$ substituents	γ-Dimethylaminopropyl-trimethoxysilane

Coupling agents are components of the finish, which forms a thin film on the glass surface. The finish contains, along with the coupling agents, a film-forming agent (binder) and the finish remains on the glass fibers. During this process trialkoxysilane is hydrolyzed to the silanol form, which is deposited on the glass surface and then immobilized [27]:

It is plausible that a slow elimination of water occurs as a result of a reaction between the silanol groups from the coupling agent and the glass [28]. Surface Si-O-Si groups are not very stable, especially in the presence of water and acidic catalysts [29].

The adhesion of silane to the glass surface is not very strong. But it has been demonstrated that using water does not completely remove aminopropyl silane from a glass surface [30].

Fillers that are surface coated with vinyl γ-methacryloxypropyl silane behave (externally) as vinyl monomers which act as additional monomers during the copolymerization of UP resins. The group consisting of $CH_2 = CH - Si =$ does not copolymerize easily [31]. On the other hand, the γ-methacryloxypropyl group copolymerizes readily [32]. Fibrous fillers are not the only compounds that are capable of combining polymer with filler by chemical bonds. Fillers in particle form behave similarly.

Bibliography to Section 6.4

[1] *Demmler, K., K. Bergmann, E. Schuch:* Kunststoffe 62 (1972), S. 845.

[2] *Hamann, D. K., W. Funke, H. Gilch:* Angew. Chem. 71 (1959), S. 596.

[3] *Boenig, H. V.:* »Unsaturated Polyesters: Structure and Properties«, Elsevier Publishing Comp., Amsterdam-London-New York, 1964.

[4] *Vieweg, R., L. Goerden:* »Polyester«, Kunststoff-Handbuch, Bd. VIII, Carl Hanser Verlag, München, 1973.

[5] *Selden, P. H.:* »Glasfaserverstärkte Kunststoffe«, Springer Verlag, Heidelberg-Berlin-New York, 1967.

[6] *Hagen, H.:* »Glasfaserverstärkte Kunststoffe«, Springer Verlag, Berlin-Heidelberg, 1956.

[7] *Doyle, E. N.:* »The Development and Use of Polyester Products«, Mc Graw-Hill Books Company, London-New York-San Francisco-Sydney-Toronto, 1969.

[8] *Lawrence, J. R.:* »Polyester Resins«, Reinhold Publishing Corp., New York, 1960.

[9] *Parkyn, B., F. Lamb, B. V. Clifton:* »Polyesters«, Bd. 2, Iliffe Books Ltd., London, American Elsevier Publishing Comp. Inc., New York, 1967.

[10] *Srna, Ch.:* In: R. Vieweg; L. Goerden: »Polyester«, Kunststoff-Handbuch, Bd. VIII, Carl Hanser Verlag, München, 1973, S. 304.

[11] *Funke, W., H. Janssen:* Makromol. Chem. 50 (1961), S. 188.

[12] *Funke, W., S. Knödler, R. Feinauer:* Makromol. Chem. 49 (1961), S. 52.
[13] *Swern, D.:* »Organic Peroxides«, Bd. 1, Interscience Publishers, J. Wiley & Sons, New York-London, 1962.
[14] *Demmler, K., J. Schlag:* Kunststoffe 64 (1974), S. 78.
[15] *Demmler, K., J. Schlag:* Kunststoffe 57 (1967), S. 566.
[16] *Bischof, C.:* Plaste Kautsch. 19 (1972), S. 936.
[17] *Pogany, G., J. Vancso-Szmercsanyi:* Plaste Kautsch. 26 (1979), S. 152.
[18] *Nicolaus, W.:* Plast. Verarbeiter 30 (1979), S. 653.
[19] *Oster, G.:* In: »Encyclopedia of Polymer, Science and Technology«, Bd. 10, Interscience Publishers, London-New York-Sydney-Toronto.
[20] *Rohmer, H., K. Heidel, F. Stürzenhofecker:* Angew. Makromol. Chem. 34 (1973), S. 71.
[21] *Roth, E.:* Dissertation at the Institut für Kunststoffverarbeitung, RWTH-Aachen, 1977.
[22] *Schulz, G. V., E. Husemann:* Z. physik. Chem. 39 (1939), S. 246.
[23] *Demmler, K., H. Lawonn:* Kunststoffe 60 (1970), S. 954.
[24] *Berger, S. E., D. R. Carr, A. J. Kane, G. S. Wooster:* Amer. Paint J. 46 (1952), S. 74.
[25] *Bovey, F. A., J. M. Kolthoff:* J. Amer. Chem. Soc. 69 (1947), S. 2143.
[26] *Ishida, H., J. L. Koenig:* Polym. Eng. and Sci. 18 (1978), S. 128.
[27] *Clark, H. A., E. P. Plueddemann:* Mod. Plast. 40 (1963), S. 133.
[28] *Lee, L. H.:* Proc. 23rd Ann. Tech. Conf. Reinforced Plast. Div., SPI, Sec. 9-D (1968).
[29] *Osthoff, R. C., A. M. Bueche, W. T. Grubb:* J. Amer. Chem. Soc. 76 (1954), S. 4659.
[30] *Schrader, M. E., A. Block:* J. Polym. Sci., C. 34 (1971), S. 281.
[31] *Vanderbilt, B. M.:* Mod. Plast. 37 (1959), S. 125.
[32] *Clark, H. A., E. P. Plueddemann:* Proc. 18th. Ann. Tech. & Manag. Conf. Reinforced Plast. Div., SPI, Sec. 20-C (1963).

6.5   Reaction Casting of Diallyl Esters

The polymerizable component in allyl compounds is the allyl group.

$$CH_2=CH-CH_2-X$$

X = activating group

It is not as easily polymerized as the vinyl group, and higher temperatures are necessary. But allyl compounds are still suitable for the reaction casting process. Among the large number of allyl compounds,

- diallyl phthalate (phthalic acid diallyl ester) and
- diethylene glycol-bis (allyl carbonate)

are the most suitable components for polymerization. Both of these monomers can be used with different comonomers. A list of various polydiallyl esters is given in [1].

6.5.1   Diallyl Phthalate Resins

The chemical formula of o-diallyl phthalate is:

o-diallylphthalate
(o-phthalic acid diallylester)
b.p. 290 °C

It is a divinyl compound that undergoes free radical polymerization at higher temperatures. Usually, free radical polymerization of divinyl compounds yields a cross-linked gel at only 1% conversion. But diallyl compounds can be polymerized to a high degree of conversion without forming a gel. The gel point depends not only on the condition of the polymerization but also on the type of monomer. According to the literature [2], the gel point of iso- and terephthalic acid diallyl esters is at 15.5% and 13.4% conversion, respectively; the gel point of o-phthalic acid diallyl ester is at 21% conversion.

For the synthesis of the resin, diallyl phthalate is polymerized almost up to the gel point; the prepolymer is isolated either by removing the remaining monomer through vacuum distillation or by precipitating the polymer through addition of certain nonsolvents, for instance isopropanol. The isolated resin is a brittle, solid product with a melting point of 80–100 °C. The melting point can be controlled by selecting the proper synthesis conditions. The resin is soluble in organic solvents, such as acetone and benzene. For the most part the resins are linear polymers, with only monofunctional reaction of the diallyl phthalate. Most of the second allyl group remains in the polymer, and thus the resin still contains a high concentration of polymerizable double bonds. The resin acts as a polymerizable compound and can be used as a prepolymer.

Diallyl terephthalate   Prepolymers

This polymerization results in polymers with short chains only, since free radical transfer by cleavage of hydrogen atoms from the allylic CH_2 group yields inactive radicals which are incapable of further polymerization. A certain percentage of cyclic compounds is characteristic of the prepolymers. These cyclic compounds are formed by an intermolecular reaction of the second allyl group. Therefore, the prepolymer does not contain one double bond per incorporated monomer unit but much less than the theoretical value. 70–75% of the volume shrinkage takes place during the preliminary polymerization. The same is true for the heat of reaction; approximately 75% of it is released during the preliminary process. Consequently, during the curing process, both the shrinkage and heat of reaction are small.

After blending with additives and fillers, the polymerization of polydiallyl phthalate prepolymers can be continued by using peroxides as a free radical source. Because of their low molecular weight, these resins melt at approximately 80–100 °C and have a low viscosi-

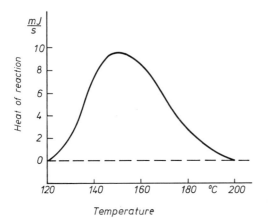

Figure 14
Differential thermoanalytical tnalysis of the curing of diallyl ahthalate resin containing 3% (by weight) of tert. butyl perbenzoate [4].

ty. This makes them suitable to be processed as casting resins and molding compounds; just as polyester resins can be reinforced with glass fibers, so can these resins; in comparison with polyester resins they have the advantage of being free-flowing. The curing reaction can be controlled by type and concentration of the radical initiator. The temperature range for the curing process of the resin is 120–200 °C using 3% by weight tert. butyl perbenzoate (Fig. 14) as radical initiator. The rate of the curing reaction is first order [3, 4]. Just as for UP resins, activators (amines or redox catalysts) can accelerate the curing process for these resins. Shrinkage caused by reaction is minimal; this is a distinct advantage of the polydiallyl phthalates. Polymerization of phthalic acid diallyl ester is accompanied by shrinkage of 11.8%; this can be further reduced to 3% by the use of prepolymers. Therefore, polydiallyl phthalate resins show low shrinkage in comparison with other resins. For the curing of unsaturated polyester resins, polydiallyl phthalate polymers can be used as monomers. The copolymerization of allyl esters and UP resin double bonds is predominantly alternating [5].

6.5.2   Polydiethylene Glycol-bis(allyl carbonate)

Diethylene glycol-bis(allyl carbonate) has the chemical formula:

$$CH_2\!=\!CH\!-\!CH_2\!-\!O\!-\!\overset{\displaystyle O}{\overset{\|}{C}}\!-\!O\!-\!CH_2\!-\!CH_2\!-\!O\!-\!CH_2\!-\!CH_2\!-\!O\!-\!\overset{\displaystyle O}{\overset{\|}{C}}\!-\!O\!-\!CH_2\!-\!CH\!=\!CH_2$$

m. p. −4 °C; b. p. 160 °C

It is also known under the trade name CR-39 (product of PPG). Free radical polymerization of the compound results in a transparent polymer with a high quality surface. It is used for manufacturing optical lenses, and as a protective coating for instruments and goggles [6].

Polymerization takes place in the mold. The casting process shows similarities to the reaction casting of PMMA. Shrinkage during the process is 14%; this is very high and has to be taken into consideration. Variable mold inserts are used. Their small walls are sealed with flexible PVC gaskets, and their larger walls are movable for changing the size of the mold.

The monomer is polymerized by free radical polymerization. Formation of a cross-linked gel occurs at relatively low conversion. As in the reaction casting of PMMA, polymerization has to proceed very slowly to avoid molding defects in the polymer. The peroxide employed determines the reaction temperature. 3% Benzoyl peroxide needs a starting temperature of 60 °C; 3% diisopropyl peroxide carbonate uses a starting temperature of 20–50 °C. The time needed for polymerization is dependent on the temperature and thickness of the molded parts. The Trommsdorff effect is less prominent during polymerization of diallyl compounds than during reaction casting of PMMA. A certain percentage of monomer remains in the final product; it is removed by tempering the part.

Since the monomer contains two polymerizable double bonds, the resulting polymers are strongly cross-linked and cannot be thermoplastically processed any further.

Polymer properties can be changed by copolymerization of diethylene glycol-bis(allyl carbonates) with various mono- or bifunctional comonomers. Methacrylic acid esters or maleic acid diesters are suitable comonomers. Elastic articles can be produced by copolymerization with butyl acrylate [7]. Homopolymers and copolymers are used only for specialty products, such as optical products. Therefore the optical industry is the main processor.

Bibliography to Section 6.5

[1] *Schildknecht, C. E.:* »Allyl Compounds and their Polymers«, High Polymers, Bd. 28, John Wiley & Sons, Inc. New York, 1973.
[2] *Lorkowski, H. J., K. Pfeiffer:* Plaste Kautsch. 22 (1975), S. 945–949.
[3] *Willard, P. E.:* SPE Journal 29 (1973), S. 38–42.
[4] *Sundstrom, D. W., M. F. English:* Polym. Engin. & Sci. 18 (1978), S. 728–733.
[5] *Pfeiffer, K., H. J. Lorkowski:* Plaste Kautsch. 26 (1979), S. 67–69.
[6] *Pechukas, A., F. Strain, W. Dial:* Mod. Plast. 20 (1943), S. 520.
[7] *Lorkowski, H. J., L. Wigant:* Plaste Kautsch. 20 (1973), S. 818–822.

6.6 Plastics of High Heat Distortion Temperature

Plastics of high heat distortion temperature are polymers which maintain good end-use properties even at high temperatures. Table 24 lists several thermoplastically processed polymers. The aerospace program and the electrical industry need lightweight polymers that withstand even higher temperatures than the polymers listed in table 24.

Polymers which melt at very high temperatures or which are cross-linked cannot be processed in the melt. Articles or semi-finished parts made from extremely high-temperature resistant polymers must be processed from intermediate polymers. These intermediate products are molded and then converted to the thermally stable polymer by a chemical reaction.

Applications for these polymers include insulating for wire and dielectric, and laminating resins, especially in combination with carbon fibers. Quantities used for these applications are small. Nevertheless, quite a bit of literature is available for this area of polymer research and processing [1–25]. In the last few years a multitude of new high temperature

Table 24   Melt processable polymers of high heat distortion temperature

Polymer	Maximum temperature for continuous use (°C)	Processing temperature (°C)
Perfluoroalkoxy-copolymer, PFA	260	320–400
Ethylene tetrafluoroethylene-copolymer, ETFE	≥ 150	260–320
Polyvinylidene fluoride	≥ 150	200–290
Polychlorotrifluoroethylene, PCTFE	170–180	260–320
Polyethersulfone	180	310–390
Polysulfone (Udel)	150	330–400
Polyarylsulfone (Astrel)	270	400
Polyphenylene sulfide	230	330–370

resistant polymers have become available, but only a small percentage of these polymers are used commercially. Only the basic types of these commercially available products are discussed in the following section.

6.6.1   Reactive Structures

In the polymer processing plant, only a few of the multitude of possible reactions that can result in high temperature resistant polymers are used. In principle the polymer processor is limited to three relevant reactions:

(a) Condensation reactions of amide and carboxylic groups that result in cyclic and cross-linked compounds

(b) Thermal polymerization of compounds with terminal acetylene groups, maleic imide groups or terminal cycloaliphatic double bonds

(c) Addition reactions to terminal maleic imide groups resulting in chain extension or cross-linkage of the macromolecule.

The synthesis of polyimides from polyamido carboxylic acid intermediates is a condensation reaction that results in cyclic and cross-linked polymers. This reaction is mainly an intramolecular condensation reaction; a heterocyclic chain unit is formed that increases dimensional stability of the intermediate at high temperatures.

Polymerization takes place during the curing of bismaleic imides. Bismaleic imides can be described as polymerizing monomers with two polymerizable double bonds per molecule. Table 25 lists several other compounds that can be polymerized further and are intermediates in the processing of high-temperature resistant resins.

Polymerization by contact molding occurs without forming any by-products and produces parts and articles that are free of molding defects, while condensation polymerization results in a volatile by-product.

Polymer producers use poly-addition of bifunctional compounds to the double bonds of bismaleic imides to synthesize monomers with long chains (intermediates). The same type of syntheses can be applied to the synthesis of high molecular weight products in situ. Besides diamines, other compounds can be used as chain extending components. Mechanism of polymer synthesis using polyaddition.

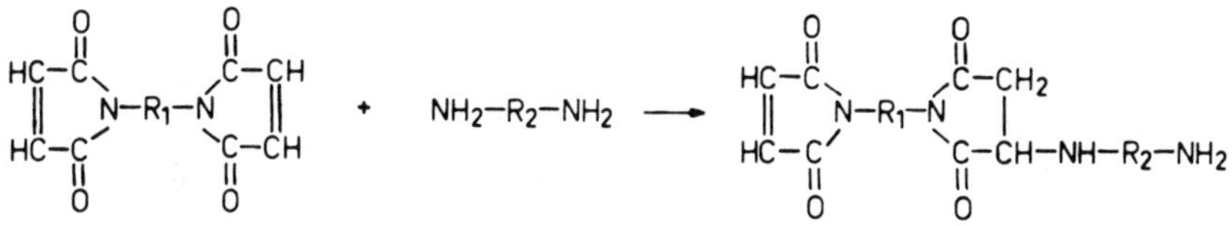

Scheme for synthesis of polymers by polyaddition

Table 25 Polymerizable structures in high temperature polymers

Chemical structure	Name	Commercial products
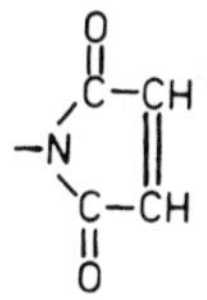	Maleic imide group	Kinel, Kerimid (Rhône Poulenc)
$-C\equiv CH$	Acetylene group	HR 600, HR 700 (Hughes Aircraft Co.)
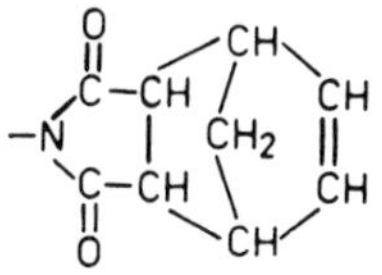	Cycloaliphatic double bond	P 13 N (Ciba Geigy AG)

6.6.2   Polyimides

The best known representatives of polymers having high heat distortion temperatures are polyimides of the general form:

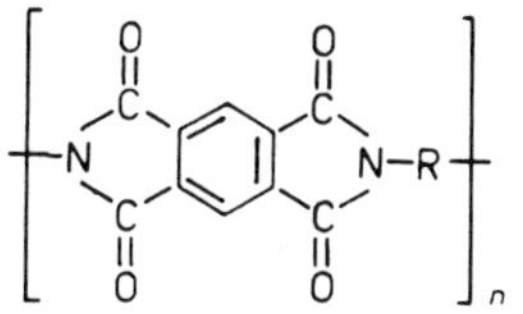

The softening point of these polymers is above 250 °C. This requires special processing methods. Processing of polyimides is done either by using highly polar solvents with very high boiling points such as dimethyl formamide, N-methyl pyrrolidone, dimethyl sulfoxide, and hexamethyl phosphoric acid amide or use of intermediates (molding is done before polymerization). Dianhydrides and diamines are basic compounds for the synthesis of polyimides. Addition of an amino group onto an anhydride group will result in a linear, thermoplastic polyamide acid which, at higher temperatures, can then be converted to a linear polyimide by intramolecular condensation. Intramolecular ring-forming reaction and intramolecular cross-linking condensation proceed side by side:

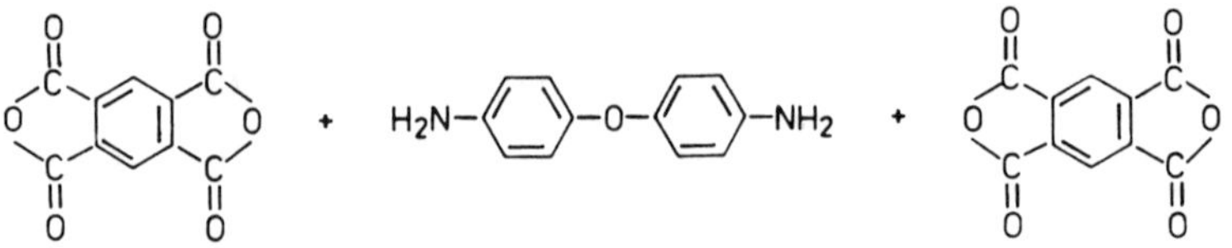

Pyromellitic
dianhydride

4,4′-Diaminodiphenyl ether

$-2 H_2O$

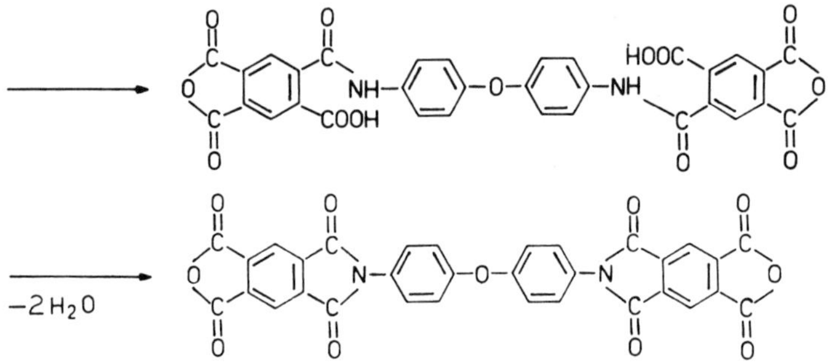

Dianhydrides with the general chemical structure:

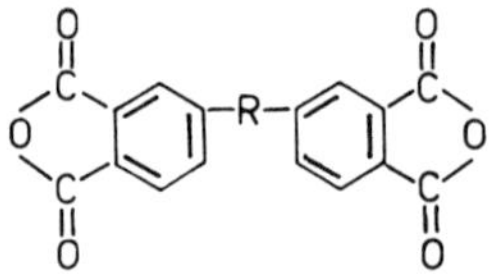

R must be a thermally stable group; for instance, $-\overset{O}{\underset{}{C}}-$ or $-\overset{CF_3}{\underset{CF_3}{C}}-$ can be used in addition to polyimides based on pyromellitic acid anhydride.

Example: For synthesis of QX 13 (commercial product of ICI), benzophenone tetra-carboxylic acid anhydride and diacetyl p,p′-diaminodiphenyl ether are prepolymerized at 250 °C in the melt. The resulting compound can be cured at 210 to 270 °C; a solution of it can be used as resin varnish for impregnating. The by-product liberated by the polyconden-sation is acetic acid:

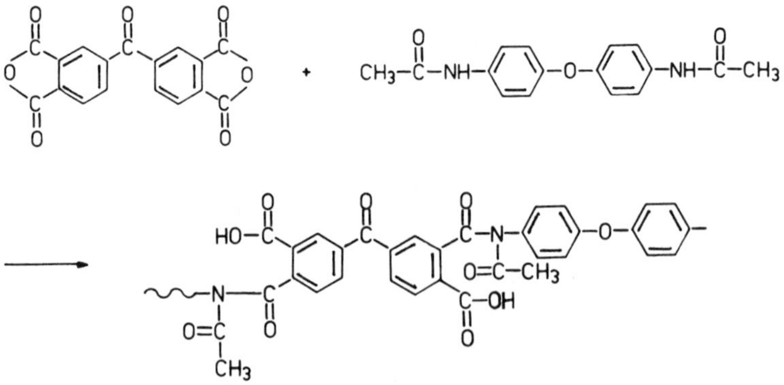

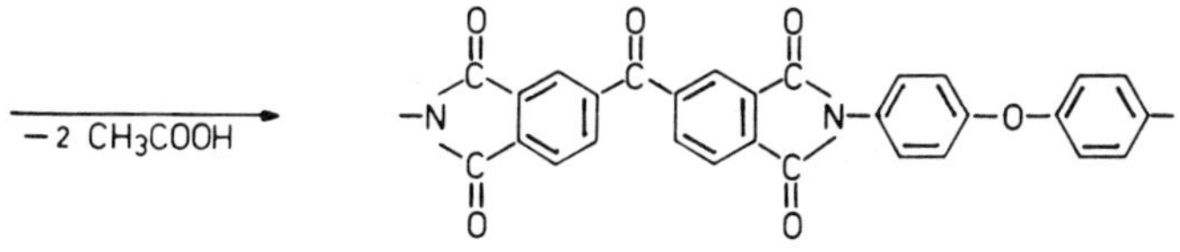

$-2 \ CH_3COOH \longrightarrow$

Polyimide QX 13 (ICI)

6.6.3 Polybismaleic Imide

Compounds with the general chemical formula

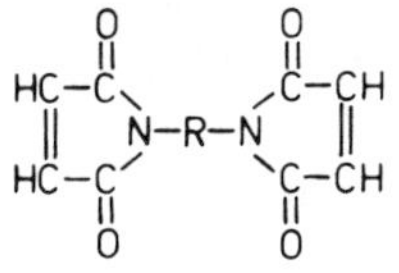

are used for polymerization. The different compounds vary in the structure of R.

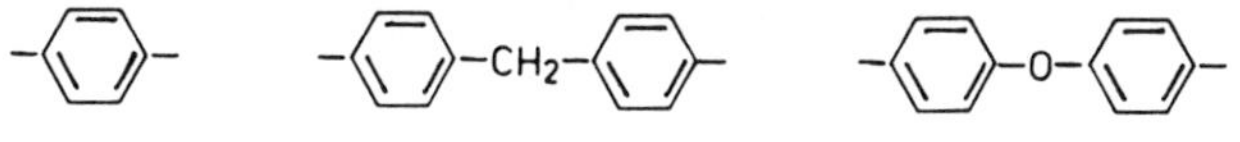

Kerimid 353 (commercial product of Rhone-Poulenc) belongs to the class of these base materials. Addition polymerization of low molecular weight bismaleic imides will yield higher molecular weight, polymerizable components.

For instance, addition reactions onto diamines will yield high molecular weight compounds:

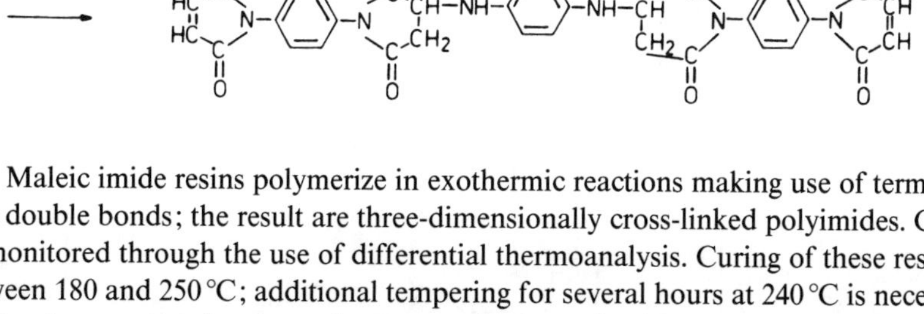

$2 \quad \cdots \quad + \quad NH_2 - \cdots - NH_2$

Maleic imide resins polymerize in exothermic reactions making use of terminal ethylenic double bonds; the result are three-dimensionally cross-linked polyimides. Curing can be monitored through the use of differential thermoanalysis. Curing of these resins occurs between 180 and 250 °C; additional tempering for several hours at 240 °C is necessary. The starting temperature for the curing process is determined by the structure of R and varies only slightly from resin to resin. For the polymerization of the following resin, the heat of polymerization is 96 kJ/mole according to [26]:

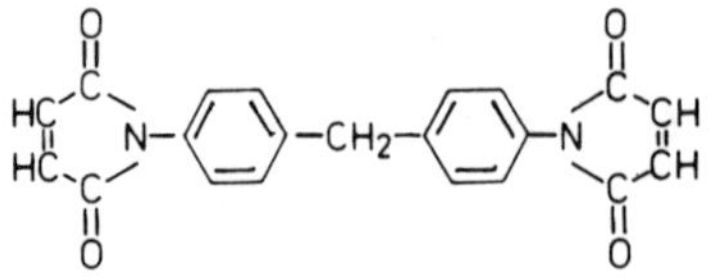

4,4'-Bismaleic imide diphenylmethane

The heat of reaction does not present any problems, especially since the molecular weights of the monomers are quite high and the heat is liberated over a lengthy nonadiabatic curing process. The size of R is inversely proportional to the number of polymerizable double bonds per unit weight.

While the heat of polymerization can easily be dissipated during the molding process, volume contraction caused by the chemical reaction is independent of the polymerization conditions and is directly proportional to the molar amount of polymerizable double bonds. Polymerizable components with high molecular weight offer an advantage. Volume contraction of bismaleic imide resin H 559 (Technochemic GmbH-Verfahrenstechnik, Heidelberg) is 4.5 vol%; shrinkage of resin H 795 is 2.1 vol% [26].

Low shrinkage on polymerization is especially important for resins used for laminating. Therefore, long-chained bismaleic imides present a distinct advantage.

6.6.4   Polyimide P 13 N

Polyimide P 13 N (commercial product of Ciba Geigy AG) is a modified bismaleic imide resin; terminal maleic imide double bonds are replaced by terminal endomethylene tetrahydrophthalic acid units which also have polymerizable double bonds. Synthesis and curing of the resin is very similar to the bismaleic imide process.

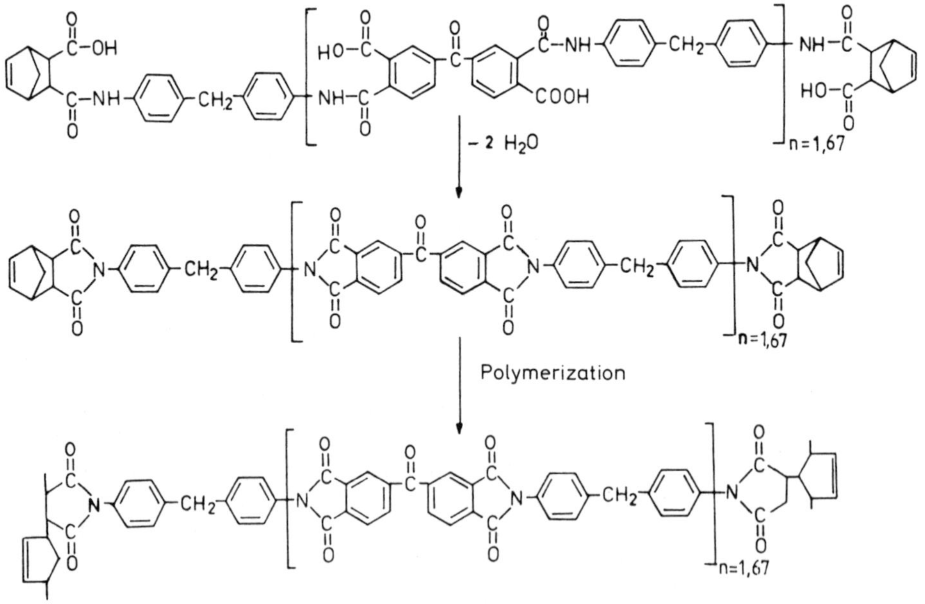

Chemistry of polyimide P 13 N [5]

The resin P 13 N is an amido carboxylic acid which is converted to a polyimide at 250–260 °C; and is subsequently cross-linked by polymerization at 310 °C.

6.6.5   Polyimide Foam Plastics

Polyimides can also be used for foamed plastics. Sky Bond (trade name of Monsanto) is an expandable polyimide made from benzophenone tetracarboxylic acid dianhydride and a diamine.

The heat resistant rigid foam recently developed by Ciba Geigy AG is based on a different principle. A new polyaddition reaction of azomethines onto bismaleic imides is employed. This reaction converts thermoplastic polymers into rigid polycyclic products [27].

Starting material is a one-component system that can be processed as powder, granules, or tablets. After melting, the material starts to foam and is then cured according to the shape of the mold and the foam density (molds of 2 cm thickness and a foam density of 0.4 g/cm^3 require 1 hr curing at 160 °C, followed by an additional hour at 180 °C). Afterwards, additional curing for 4–6 hours at 180 °C has to be continued outside the mold. The reaction mechanism is:

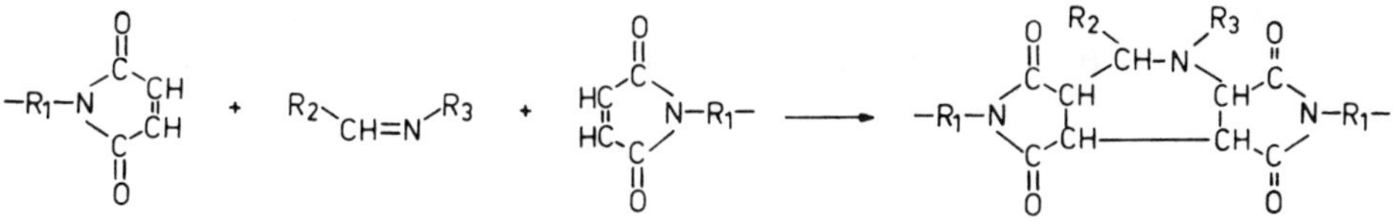

6.6.6   Resins with Acetylene as Base Material

Newly developed resins by Hughes Aircraft Co. (USA) display terminal acetylene groups which can be polymerized [28]. These products are polyimides (in principle). They can be synthesized in the following manner:

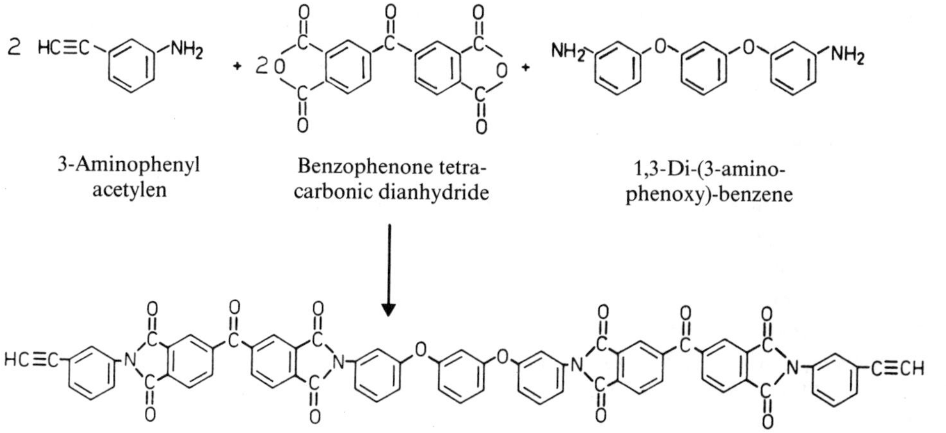

3-Aminophenyl acetylen	Benzophenone tetra- carbonic dianhydride	1,3-Di-(3-amino- phenoxy)-benzene

The resins are cured at temperatures which are above their softening points (approximately 220 °C or higher). Presumably, the reaction mechanism involves trimerization of the acetylene group, which results in formation of an aryl group.

Bibliography to Section 6.6

[1] *Adrova, N. A., M. I. Bessonov, L. A. Lains, A. P. Rudakov:* »Polyimides – A new class of heat resistant polymers«, Israel Program for Scientific Translation, Jerusalem, 1969.

[2] *Behr, E.:* Hochtemperaturbeständige Kunststoffe, C. Hanser Verlag, München, 1969.

[3] *Korshak, V. V.:* »Heat-Resistant Polymers«, Israel Program for Scientific Translations, Jerusalem, 1971.

[4] *v. Krevelen, D. W.:* »Properties of Polymers, Correlations with chemical structure«, Elsevier Publishing Comp., Amsterdam-London-New York 1972.

[5] *Elias, H. G.:* »Neue polymere Werkstoffe 1969-1974«, C. Hanser Verlag, München, 1975.

[6] *Ziegmann, G.:* Dissertation at the Institut für Kunststoffverarbeitung, RWTH Aachen, 1979.

[7] *Behr, E.:* Kunststoffe 62 (1972), S. 670.

[8] *Berlin, A. A.:* Plaste Kautsch. 21 (1974), S. 486.

[9] *Deneke, W. H.:* Industrieanzeiger 91 (1969), S. 1127.

[10] *Goedtke, P., H. Mathes, W. Schäfer:* Kunststoffe 68 (1978), S. 687.

[11] *Kataoka, S.:* Current Topics of Polyimide Type Heat Resistant Resins Jpn. Plast. Age (1977), S. 331.

[12] *Kuriatkowski, G. T., L. M. Robeson, A. W. Bedwin:* J. Polym. Sci. 13 (1975), S. 961.

[13] *Lauer, W.:* Plastverarbeiter 27 (1976), S. 305.

[14] *Lucke, H.:* Kunststoff-Rundschau 18 (1971), S. 434.

[15] *Meckelburg, E.:* Gummi + Asbest Kunststoffe 27 (1974), S. 414.

[16] *Merten, R.:* Angew. Chemie 83 (1971), S. 339.

[17] *Merten, R.:* Synthesis of Heterocyclic Ring Systems for Heat-Resistant-Plastics from Polyisocyanates in: Advances in Urethane-Science and Technology Volume 2, Technomic Publ. Co. Inc. Westport, Conn., USA 1973, S. 123-140.

[18] *N. N.:* Kunststoffe 67 (1977), S. 17.

[19] *N. N.:* Gummi, Asbest, Kunstst. 29 (1976), S. 432.

[20] *Russo, M.:* Kunststoffe 65 (1975), S. 346.

[21] *Sokolov, L. B.:* Jpn. Plast. Age 13 (1975), S. 47.

[22] *Stenzenberger, H.:* J. Appl. Polym. Sci., Applied Polymer Symposium 31 (1977), S. 91.

[23] *Stenzenberger, H.:* Kautsch. Gummi, Kunstst. 29 (1976), S. 477.

[24] *Wolfer, W.:* Gummi, Asbest, Kunstst. 25 (1972), S. 92.

[25] *Reese, E.:* Kunststoffe 62 (1972), S. 733.
[26] *Stenzenberger, H. D.:* »Hochtemperaturbeständige Kohlenfaserprepregs«, Final report for the research project Nr. F 70-95068/10, Dez. 1978.
[27] *Schmitter, A.:* »Eigenschaften eines neuen wärmebeständigen Hartschaumes«, Discours at the 6. Intern. Schaumstoff-Tagung, May 1976 in Düsseldorf.
[28] *Bilow, N., A. L. Landes, T. J. Aponyi:* Thermosetting Acetylene-Substituted Polyimides, Firmenschrift der Hughes Aircraft Comp., Culver City, Col., USA.

6.7 Reaction Casting of Polyamides

6.7.1 General

Polyamides belong to the class of thermoplastic polymers where the amide group is the structure determining group.

$$- \overset{\text{O}}{\underset{\text{||}}{\text{C}}} - \text{NH} -$$ Amide group

Polyamides are synthesized from diamines and dicarboxylic acids or from amino carboxylic acids, specifically their anhydrides (lactams). The following base components are used for the synthesis of the most important polyamides:

$HOOC - (CH_2)_4 - COOH$ Adipic acid

$H_2N - (CH_2)_6 - NH_2$ Hexamethylenediamine

ε-Caprolactam

Polyamide 6.6 is made by polycondensation of adipic acid and hexamethylenediamine. Polymerization of ε-caprolactam will yield polyamide 6.

Polymer synthesis using lactams is free of by-products and can therefore be classified as a polymerization. Occasionally, the polymerization of lactams is classified as a polycondensation. This is justified, since some steps of the hydrolytically initiated polymerization are condensation reactions.

Two different methods may be used for the synthesis of polyamide 6 from ε-caprolactam: hydrolytic polymerization and activated anionic polymerization.

The technical synthesis of granular polyamide makes use of hydrolytic polymerization. This reaction is described later, but reaction casting of polyamide 6 is usually based on anionic activated polymerization. While hydrolytic polymerization takes several hours, anionic activated polymerization is completed in less than a few minutes. Therefore, the anionic polymerization is of more interest to the processor. A description of both reactions follows.

A summary of the synthesis, processing, and properties of polyamides is given in [1].

6.7.2   Hydrolytic Polymerization of ε-Caprolactam

Polyamide 6 is formed by heating aqueous (0.01–1 mole%) ε-caprolactam (m.p. 70_C) above 200 °C. Water reacts with ε-caprolactam and forms ε-aminocaproic acid during a preliminary step.

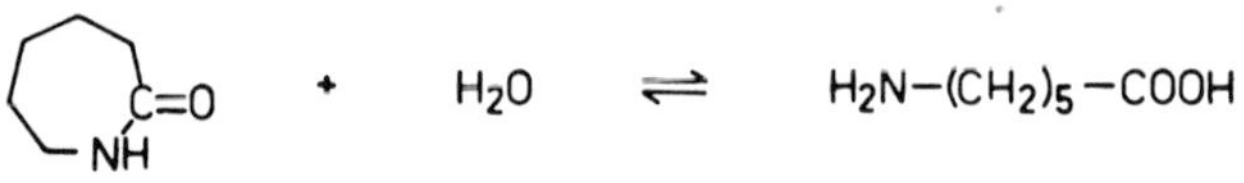

$$\text{[ε-caprolactam ring] C=O, NH} \quad + \quad H_2O \quad \rightleftharpoons \quad H_2N-(CH_2)_5-COOH \tag{1}$$

ε-caprolactam +   Water $\rightleftharpoons$ ε-aminocaproic acid

The actual polymerization initiator is the newly formed aminocarboxylic acid. Two different mechanisms are possible for chain propagation:

– Condensation between two ε-aminocaproic acid molecules; or condensation between amino and carboxylic groups available in free amino acid and oligomer/polymer intermediate.

– Addition of an amino group belonging to a free amino acid to ε-caprolactam.

The hydrolysis of ε-caprolactam is a very slow reaction. Compounds with available amino groups such as ε-aminocaproic acid are able to initiate the reaction much faster; the carboxylic group acts as an additional accelerator [2].

The addition reaction is the faster of the two secondary reactions:

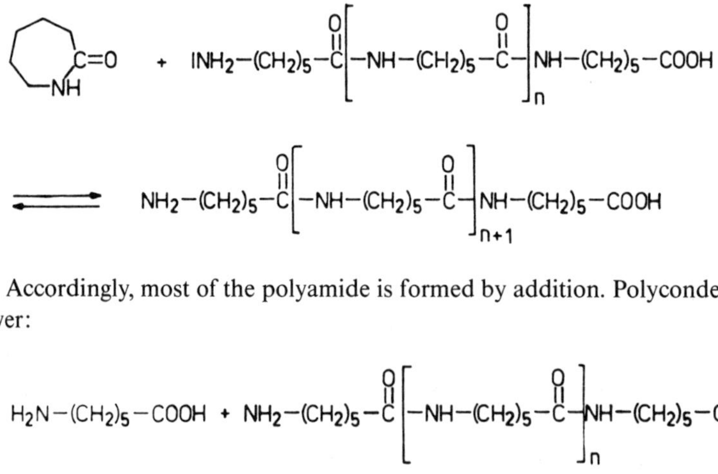

$$\text{[ε-caprolactam ring] C=O, NH} \quad + \quad NH_2-(CH_2)_5-\overset{O}{\underset{\|}{C}}\left[-NH-(CH_2)_5-\overset{O}{\underset{\|}{C}}-\right]_n NH-(CH_2)_5-COOH \tag{2}$$

$$\rightleftharpoons \quad NH_2-(CH_2)_5-\overset{O}{\underset{\|}{C}}\left[-NH-(CH_2)_5-\overset{O}{\underset{\|}{C}}-\right]_{n+1} NH-(CH_2)_5-COOH$$

Accordingly, most of the polyamide is formed by addition. Polycondensation is much slower:

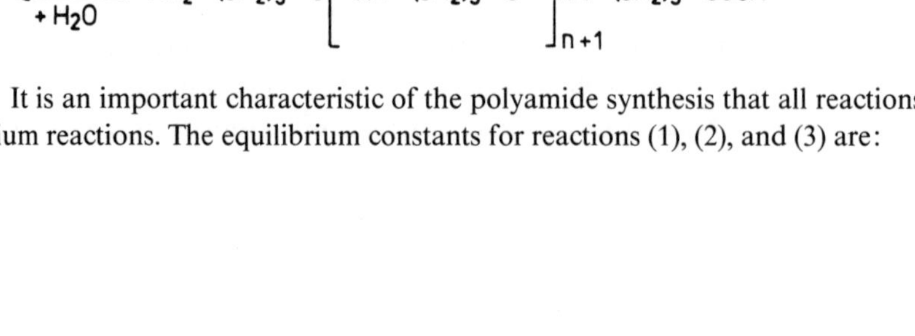

$$H_2N-(CH_2)_5-COOH \; + \; NH_2-(CH_2)_5-\overset{O}{\underset{\|}{C}}\left[-NH-(CH_2)_5-\overset{O}{\underset{\|}{C}}-\right]_n NH-(CH_2)_5-COOH \tag{3}$$

$$\underset{+H_2O}{\overset{-H_2O}{\rightleftharpoons}} \quad NH_2-(CH_2)_5-\overset{O}{\underset{\|}{C}}\left[-NH-(CH_2)_5-\overset{O}{\underset{\|}{C}}-\right]_{n+1} NH-(CH_2)_5-COOH$$

It is an important characteristic of the polyamide synthesis that all reactions are equilibrium reactions. The equilibrium constants for reactions (1), (2), and (3) are:

$$K_1 = \frac{[(\text{Aminocapr.})_1]}{[\text{Caprol.}] \cdot [\text{H}_2\text{O}]} \tag{4a}$$

$$K_2 = \frac{[(\text{Aminocapr.})_{n+1}]}{[\text{Caprol.}] \cdot [(\text{Aminocapr.})_n]} \tag{4b}$$

$$K_3 = \frac{[(\text{Aminocapr.})_{n+1}] \cdot [\text{H}_2\text{O}]}{[(\text{Aminocapr.})_1] \, [(\text{Aminocapr.})_n]} \tag{4c}$$

Caprol. $= \varepsilon$-caprolactam
$(\text{Aminocapr.})_1$ $= \varepsilon$-aminocaproic acid
$(\text{Aminocapr.})_n$ $=$ polycondensate from n monomer molecules
K_1 to K_2 $=$ equilibrium constant
[ ] $=$ notation for concentration

According to the equilibrium constants, and independent of the type of synthesis used, the components (polymers, oligomers, monomers, and water) exist in certain mass ratios in a thermodynamic equilibrium. At normal processing temperatures, polyamide is not affected by water, but in order to avoid loss of molecular weight caused by hydrolysis, the water must be removed by drying prior to the thermoplastic processing.

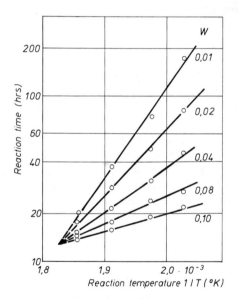

Figure 15
Time needed to reach condensation equilibrium during polymerization of caprolactam as a function of temperature T (K) at various amounts of initial water content W (mole/mole) [3].

The rate of polymerization depends on the temperature and on the percentage of water being produced (Figure 15). Polymerization continues until an equilibrium is reached. This procedure is especially necessary for the reaction casting process in order to retain only small amounts of monomer in the end product, since any monomer remaining is almost impossible to remove from the finished part. It is not possible to shorten the lengthy polymerization time by increasing the amount of water as initiator, since this would affect the degree of polymerization [3]. The average degree of polymerization, P_n, decreases as the water in-

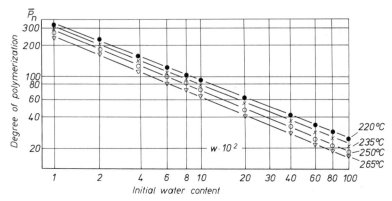

Figure 16   Degrees of polymerization P_n at equilibrium as a function of initial water content W (mole/mole) in caprolactam at various temperatures [3].

creases (Figure 16). In order to achieve the high degree of polymerization necessary for high polymer quality, the amount of water has to be kept small and allowance has to be made for long polymerization times.

6.7.3   Reaction Casting involving Hydrolytic Polymerization

Hydrolytic polymerization at ambient pressure is used to manufacture large blocks, especially cylinders of diameters up to 500 mm, from which large equipment parts can be made by machining [1]. For this reaction the molten ε-caprolactam (m.p. 70 °C) is mixed with 2–5% of a salt made from adipic acid and hexamethylenediamine (AH salt) ε-aminocaproic acid or with 0.6% water; it is then heated to 275 °C and poured into heated metal molds. Equilibrium occurs after 24 hours of reaction time. Monomer content is 9–12%.

Few problems result from the heat of polymerization, since it is low (125 kJ/kg) and is dissipated over a long time span. Also, the volume contraction caused by reaction is small, 2–3 vol % (difference in density between ε-caprolactam and melt at 20 °C). The shrinkage caused by crystallization and thermal contraction is substantially higher. To avoid thermal stress inside large parts, the molten polycondensate has to be cooled very slowly. The polymer is coarse textured.

6.7.4   Activated Anionic Polymerization

A mixture of sodium caprolactam and ε-caprolactam polymerizes very quickly if heated higher than 190 °C. A high concentration of sodium caprolactam will complete the anionic polymerization in less than 1 minute [4].

Sodium caprolactam is very sensitive to water; therefore, it cannot be formed by reacting aqueous NaOH with ε-caprolactam. The reaction of ε-caprolactam with benzyl sodium followed by evaporation of toluene, the by-product, represents a very convenient synthesis.

The polymerization is initiated by addition of the ionic initiator onto ε-caprolactam (Eq. 5):

$$\text{(ring with } C=O, N^- Na^+) + \text{(ring with } HN, O=C) \longrightarrow \text{(ring with } C=O, N-C=O, (CH_2)_5-NH^- Na^+) \qquad (5)$$

This primary step results in formation of a reactive sodium-salt which will also react with available ε-caprolactam (Eq. 6).

$$\text{(ring } C=O, N-C(CH_2)_5-NH^- Na^+) + \text{(ring } C=O, NH) \longrightarrow \text{(ring } C=O, N^- Na^+) + \text{(ring } C=O, N-C-(CH_2)_5-NH_2) \qquad (6)$$

Subsequently, the initiator (sodium lactam) is restored, and the newly formed acyllactam, a stable compound, reacts very quickly with sodium lactam (Eq. 7).

$$\text{(ring } C=O, N^- Na^+) + \text{(ring } C=O, N-C-(CH_2)_5-NH_2) \longrightarrow \text{(ring } C=O, N-C-(CH_2)_5-N-C-(CH_2)_5-NH_2, Na^+) \qquad (7)$$

Chain propagation occurs in several steps: The sodium salt that is formed reacts immediately with ε-caprolactam and regenerates sodium lactam, the initiator, which in turn will react with the acylated lactam and produce a growing polymer molecule.

Neither the reaction of acyllactam with sodium lactam, nor the other steps of the chain propagation reaction require higher temperatures. For the ionic polymerization of lactam with sodium lactam as the only initiator, a starting temperatures of more than 190 °C is necessary. This temperature can be reduced to 100 °C if acyllactam is added to the reaction mixture. Acyllactam "activates" the polymerization. The process is called "activated anionic polymerization".

ε-Caprolactam which has been acylated by ε-aminocaproic acid is not the only available activator. The essential characteristic of acyllactam is the activation of the cyclic nitrogen atom by the adjacent carbonyl groups. This also makes several other acyllactams suitable activators.

Of great commercial interest is the rapid rate of polymerization at relatively low temperatures made possible by the simultaneous use of sodium lactam and acyllactam. The quality of the processed parts is reproducible if pure components are used. The heat of reaction is small compared to the heat of polymerization of styrene and methyl methacrylate and does not present any problems. The polymerization can go directly from molten ε-caprolactam to a solid article.

The molecular weight of the polymerizate can be controlled by the concentration of acyllactam as long as the bulk temperature does not reach more than 190 °C. At higher temperatures sodium lactam reacts with ε-caprolactam according to equation (5). Additional acyllactam is formed and the concentration of molecules propagating new chains increases. This results in a polymer with a lower molecular weight.

Acyllactam is also unstable in the presence of water, and therefore only compounds that are absolutely free of water should be used for the synthesis and humidity in the air must be avoided during storage.

A suitable synthesis of acyl compounds is the reaction of ε-caprolactam with iso-cyanate [5]. In a preliminary step, some of the isocyanate removes the available water by reacting with it. The remaining isocyanate reacts with ε-caprolactam:

$$H_2O \quad + \quad 2\ R-N=C=O \quad \longrightarrow \quad CO_2 \quad + \quad R-NH-\overset{\overset{\textstyle O}{\|}}{C}-NH-R \tag{8}$$

$$\underset{NH}{\overset{}{\diagdown}}C=O \quad + \quad O=C=N-R \quad \longrightarrow \quad \underset{N-C-NH-R}{\overset{}{\diagdown}}C=O \tag{9}$$

Activated lactams containing acyl radicals without any carboxyl groups produce poly-mers free of carboxyl groups provided that polymerization takes place at $T < 190\,°C$.

6.7.5   Reaction Casting Using Ionic Polymerization

The processing of parts and semifinished goods made from polyamide 6 in combina-tion with an anionic activated polymerization is characterized by very short reaction times. A polymerization time of 1–5 minutes allows automated manufacturing with very short cy-cle times.

This process is applicable to the manufacture of very large parts, for example, gasoline tanks for helicopters and fuel oil tanks with volumes of 1000–3000 L [6]. In addition, reac-tion casting is used to manufacture gears, rollers, sheets, and blocks of semifinished goods.

The reaction casting process makes it necessary to use an initiator and an activator which, when separated and in the absence of humidity, are stable at room temperature; nei-ther the initiator by itself nor the activator by itself should initiate the polymerization of lac-tams ($< 190\,°C$). The reaction casting starts with a molten material. Two types of processes are in use [6]

- Two melts are mixed together in definite proportions; one melt contains the catalyst and the other the activator. The molds are filled immediately with well mixed melts.
- The molten lactam which contains the catalyst and a liquid activator (accelerator) in a mass ratio of 200:1 to 300:1 is fed in. This puts great demands on the weight feed-er and the mixer.

Water and other compounds with reactive hydrogen atoms should not be present dur-ing any of these reactions. Molten lactam which is not used immediately has to be protected against oxygen. At temperatures of 100–160 °C (depending on mold size and method), the low-viscosity lactam melt is poured into a preheated mold. After an induction period of ap-proximately 30 seconds a rapid polymerization starts, the viscosity increases, and the melt becomes turbid (Figure 17). The reaction reaches equilibrium in a few minutes. At this point, the polymer contains 1–2% monomer. Usually the product is released from the mold without cooling the mold, since the polymer is dimensionally stable at temperatures prevail-ing at the end of the reaction.

Polyamide 6 shows only limited solubility in molten ε-caprolactam and will partially crystallize during polymerization. The total heat of polymerization is 125 kJ/kg; after add-

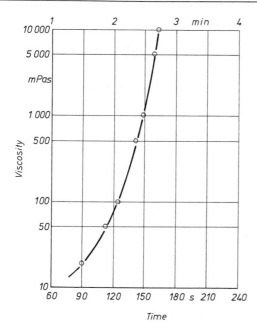

Figure 17
Change in viscosity of a polymerizing
melt of caprolactam as a function
of time [6].

ing the heat of crystallization (at 60% crystallization), the total amount of heat liberated is
238 kJ/kg. Since the average specific heat is 2.1 kJ/kg, the liberated heat will produce a temperature rise of 70–80 °C.

The volume change for the ambient pressure process is primarily 4% contraction due
to crystallization and 5–8% thermal contraction (depending on the maximum bulk temperatures). Shrinkage due to the reaction is minimal.

The activated anionic lactam polymerization can be carried out using mixtures of ε-ca-
prolactam and other lactams; an interesting variation is the use of dilactams, such as methylene biscaprolactam.

The use of dilactams results in cross-linked polyamides. The polymer processor can
vary the polymer properties by altering the monomer composition and thus is able to produce polymers suitable for specific applications. Examples of possible reaction compositions are listed in Table 26.

Table 26   Examples of formulations for the activated polymerization of caprolactam [1 c].

	a	b
ε-Caprolactam %	96–90%	81–77%
Laurin- or caprolactam %	–	6–10%
Caprolactam-Sodium	2–5%	6%
Activator % (example: N-acetyl-caprolactam)	2–5%	4,5%
Cross-linking agent % (example: methylene-bis-caprolactam)		2,5%
Polymerization temperature	130–160 °C	120–150 °C

Bibliography to Section 6.7

[1] a) *Hopff, H., A. Müller, F. Wenger:* »Die Polyamide«, Springer Verlag, Berlin-Göttingen-Heidelberg, 1954.
   b) *Klare, H., E. Fritzsche, V. Gröbe:* »Synthetische Fasern aus Polyamiden – Technologie und Chemie«, Akademie-Verlag, Berlin, 1963.
   c) *Vieweg, R., A. Müller:* »Polyamide«, Kunststoff-Handbuch, Bd. VI, Carl Hanser Verlag, München, 1966.
[2] *Heikens, D., P. H. Hermans, G. M. van der Want:* J. Polym. Sci. 44 (1960), S. 437.
[3] *Reimschüssel, H. K.:* J. Polym. Sci. 41 (1959), S. 457.
[4] *Hanford, W. E., R. M. Joyce:* J. Polym. Sci. 3 (1948), S. 167.
[5] DE-PS 1.067.591 (1956), Bayer AG.
[6] *Dhein, R., R. V. Meyer, F. Fahnler:* Kunststoffe 68 (1978), S. 2.

7 Polycondensation Reactions in the Polymer Processing Plant

7.1 General Comments on Polycondensation Reactions

A polycondensation is a chemical reaction between two or more functional molecules which results in a macromolecular compound and the elimination of low molecular weight by-products. Bifunctional low molecular compounds are necessary for forming linear macro molecules. In principle, a condensation reaction is possible between two reactive groups of the same type; for example, the cross-linking of silanol in the silane cross-linking process (see Section 9.5).

Usually polycondensation is a reaction between two different functional groups which may either exist separately in two different low molecular weight compounds or be a part of the same molecule. During polycondensation, at least three components are present in the reaction vessel the starting component, the polycondensate, and the by-product.

If the two reactive groups are members of two different starting compounds, then four components participate in the reaction.

$$X-R^1-X \quad + \quad Y-R^2-Y \quad + \quad X-R^1-X \quad + \quad Y-R^2-Y$$

$$\downarrow$$

$$X-R^1-R^2-R^1-R^2-Y \quad + \quad 3 \ X \ Y$$

X and Y are different functional groups
XY is a by-product
X = Y if condensation takes place between the equally functional groups.

Polycondensation, unlike polymerization, is a type of reaction that is not easily governed by uniform laws and rules. X and Y may have totally different characteristics, and the condensation reactions may involve different chemical reactions.

In the polymer processing plant, a polycondensation reaction may be employed if only one reaction step is necessary to convert easily manageable intermediates with good shelf lives into the finished product. Usually, the use of solvents is not possible; it is only indicated in special cases, such as in the manufacture of films and laminates. One has to consider also the type and amount of by-products. For instance, HCI and other corrosive by-products cannot be tolerated in the processing plant. Therefore, only a very few polycondensation reactions are used by the plastics processor. Processes based on polycondensation reactions are used to manufacture parts and semifinished goods made from:

- Phenol formaldehyde condensates (phenol can be replaced by other simple aromatic compounds)
- Urea formaldehyde condensates (urea can be replaced by other activated amino compounds)
- Melamine formaldehyde condensates
- Polyimide, from polyamide carboxylic acids
- Polycarbodiimide foams

- Polysiloxanes
- Cross-linked polyethylene (according to the method of silane cross-linking).

By examining the chemical mechanism of cross-linking reactions of elastomers, additional reactions can be observed whose by-products are able to react further and thus do not appear as volatile compounds.

The synthesis of polysiloxanes and the cross-linking of silane will be discussed as polymer reactions, together with other cross-linking reactions. Because of the similarity of the polycarbodiimide process to the polyurethane process, both will be discussed in the chapter that describes polyaddition reactions (Chapter 8).

Polycondensation reactions show two characteristics which are important for the polymer processor:

- They liberate volatile by-products.
- The condensation steps are mutually independent.

The volatile by-product is in most cases water, but carbon dioxide is liberated during the polycarbodiimide process. During the manufacture of solid parts, the release of a volatile product can result in voids in the polymer. Therefore, it is necessary to keep the amount of volatile compounds to a minimum. At the pressure employed in the manufacture of parts, small amounts of by-products remain dissolved. Hydrophilic fillers, which are frequently used in the manufacture of phenolics and amino resins, absorb the water formed as a by-product, thus eliminating it as a cause of voids. In order to minimize the formation of volatile products, precondensates with a suitable degree of condensation are used. During the conversion to the final product, only small amounts of by-products are liberated.

Processing polycarbodiimide from low molecular weight diisocyanates releases large amounts of carbon dioxide. The production of polycarbodiimides in the polymer processing plant is suitable for the manufacture of articles made of foam, particularly foamed semifinished goods. The carbon dioxide formed as a by-product during the reaction is utilized as a blowing agent in the manufacture of foam. Also, during the manufacture of phenol formaldehyde or urea formaldehyde foam plastics the formation of a volatile by-product does not present an interference.

Every step of a condensation reaction is a self contained reaction. This constitutes a very important difference between polymerization and polycondensation. Each condensation step has to be independently activated. While polymerization is a chain reaction, polycondensation is a step reaction. Therefore, condensation steps occur statistically and do not prefer components already partially reacted. As a result, molecular weight and viscosity increase only slowly as the degree of condensation increases. The same relation is valid for polyaddition (see Figure 18).

Prepolymerizates are easy to process, since their molecular weights are small even at a high degree of conversion. In the processing plant, polycondensation combines in combination with the molding proceeds with only minimal molar conversion of the components and therefore limited liberation of by-products; nevertheless, it facilitates a rapid increase in viscosity and accelerates the curing of the molded article.

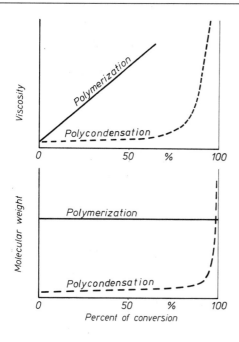

Figure 18
Schematic graph of viscosity and
molecular weight during polymerization
and polycondensation.

7.2   Phenolics and Amino Resins

Phenolics and amino resins are the most important polycondensation resins. They are used in powder or liquid form for the molding of articles, in the manufacture of foamed plastics; as a coating for paper and laminated fabrics; in the chemical treatment of textiles; as a binder for grinding disks, laminated sheets, and so on, and as an adhesive in the production of plywood, resins bonded pressed wood, chipboard etc. Regardless of the end use and form of delivery, the curing reaction is always a polycondensation reaction.

According to the many different applications, a great many types of products are used as aqueous solutions for the chemical treatment of textiles. Similarly, the manufacture of foamed amino resins starts with aqueous solutions of low molecular weight compounds. In addition to their use as aqueous solutions, these resins find applications as anhydrous, liquid products (casting resins) or as molding compounds in the form of brittle, thermoplastic, high molecular weight solids. A great many resins based on phenol formaldehyde or amine formaldehyde are commercially available, whose applications depend on the chemical composition of the resin, the amount of water contained in the raw material to be processed, the degree of condensation of the intermediates, and the presence of different kinds and amounts of fillers. The phenolics and amino resins employed by the polymer processor are rarely pure resins; generally, they are mixtures containing a large amount of filler. The fillers produce reasonably priced polymers and give the polymer specifically requested properties. It is a characteristic of the phenolics and amino resins that they produce excellent composites with most fillers. When sawdust is used as a filler, the resin will penetrate even its smallest cavity. In some instances, the resin may react chemically with the filler; for instance, resins are able to react with wool and certain components of wood.

Water formed as a by-product during condensation reactions may be detrimental to the properties of cast and molded articles. Products that contain a hydrophilic filler such as sawdust are able to absorb quite a large quantity of water, but obviously not an unlimited amount. These water-related problems often make it necessary to use precondensates with a high degree of conversion. In order for the polymer to have good processibility, low molecular weight compounds have to be used; on the other hand, only high molecular weight precondensates will be able to minimize the problems connected with the release of water.

For example, the reaction of 1 kg of an anhydrous mixture of equimolar amounts of phenol and pure formaldehyde will produce 855 g of polymer and 145 g of water. Theoretical calculations show that the molding compounds used in the polymer processing plant must have a high degree of preliminary condensation.

It is the task of the raw material manufacturer to proceed with the polycondensation reaction until the condensation product has reached the point of optimum processing properties. It is then the task of the polymer processor to continue the condensation reaction. The only case in which the liberation of large quantities of water does not present a problem is in the processing of foamed plastics.

The total tonnage of phenolics and amino resins processed is quite high. They also account for a high percentage of total polymer production. But only a small part of the total production of these resins is used for the processing of polymer parts or semifinished goods. The major portion is used for gluing applications. In the United States and Canada, only 17% of the total production of phenolic resins and 6% of the total production of amino resin were used for the manufacture of articles in the polymer processing plant during 1980

Table 27   Use of phenolic resin in Unites States and Canada during 1980 [13]

Application	Consumption (1000 t)	% of total consumption
Glues, binders	246	35
Laminating resins	44	6
Wood refining	210	30
Molding compounds	120	17
Laquers	9	1
Export	12	2
Other	60	9
Total	701	100

Table 28   Use of urea and melamine resins in United States and Canada during 1980 [13].

Application	consumption (1000 t)	% of total consumption
Glues, binders	447	76
Molding compounds	35	6
Paper processing	30	5
Protective coating	28	5
Textile processing	34	6
Export	8	1
Other	7	1
Total	589	100

(Tables 27 and 28). Detailed information is not available for the amount of phenolics and amino resins used in Europe during 1980. The Federal Republic of Germany processed 68,000 tons of these products in 1977 [1].

A large amount of literature deals with the chemistry, manufacture, and processing of phenolics and amino resins [2–12].

7.3   Common Chemical Properties of Phenolics and Amino Resins

The curing of phenolics and amino resins involves the same reactions as the synthesis of their corresponding precondensates. The synthesis of these products has to be discussed in detail to gain a clear understanding of the curing reactions.

Phenolics and amino resins are formed by reacting with an aldehyde and a compound containing several reactive hydrogen atoms. The term "reactive hydrogen atoms" means that a substitution of hydrogen atoms by aldehydes is possible.

$$R-C\underset{O}{\overset{H}{\lessgtr}} \quad \text{Aldehyde} \quad -C\underset{O}{\overset{H}{\lessgtr}} = \quad \text{Aldehyde group}$$

$$H-\bigcirc-H \quad \text{Component containing hydrogen atoms which are reactive with aldehyde}$$

The aldol condensation of acetaldehyde is an example of a simple reaction of an aldehyde group with a component containing reactive hydrogen atoms:

$$CH_3-C\underset{H}{\overset{O}{\lessgtr}} + CH_3-C\underset{H}{\overset{O}{\lessgtr}} \longrightarrow CH_3-\underset{OH}{\overset{H}{\underset{|}{C}}}-CH_2-C\underset{H}{\overset{O}{\lessgtr}}$$

Acetaldehyde   Aldol

In the presence of strong alkali, aldol will further condense to "aldehyde polymers." The formation of phenolics and amino resins follows a similar mechanism. This reaction does not occur between components of the same kind; rather, the aldehyde reacts with the reactive hydrogen atoms of a nonaldehyde component.

The primary reaction is an addition reaction:

$$H-\bigcirc-H \quad + \quad \underset{R}{\overset{H}{\underset{|}{\overset{|}{C}}}}=O \quad \longrightarrow \quad H-\bigcirc-\underset{R}{\overset{H}{\underset{|}{\overset{|}{C}}}}-OH \qquad \text{(a)}$$

Base component
with reactive   +   aldehyde → alcohol
hydrogen atoms

The high reactivity of the intermediate makes a rapid propagation reaction possible, as follows: Either two alcohols react to form an ether, or the alcohol reacts directly with the base compound.

$$H-\bigcirc-\underset{\underset{R}{|}}{\overset{\overset{H}{|}}{C}}-OH \quad + \quad HO-\underset{\underset{R}{|}}{\overset{\overset{H}{|}}{C}}-\bigcirc-H \quad \longrightarrow \quad H-\bigcirc-\underset{\underset{R}{|}}{\overset{\overset{H}{|}}{C}}-O-\underset{\underset{R}{|}}{\overset{\overset{H}{|}}{C}}-\bigcirc-H \quad + \quad H_2O \qquad (b)$$

$$H-\bigcirc-\underset{\underset{R}{|}}{\overset{\overset{H}{|}}{C}}-OH \quad + \quad H-\bigcirc-H \quad \longrightarrow \quad H-\bigcirc-\underset{\underset{R}{|}}{\overset{\overset{H}{|}}{C}}-\bigcirc-H \quad + \quad H_2O \qquad (c)$$

The conditions chosen for the reaction will determine if either reaction (b) or (c) will be the only reaction or if one will occur prior to the other.

Aldehyde, the base component, acts as a bifunctional compound and is able to react with certain components containing two or more reactive hydrogen atoms, resulting in the formation of chains or cross-linkages. Since the compounds formed in reactions (b) and (c) have additional reactive hydrogen atoms, renewed addition of an aldehyde is possible; this allows the condensation to continue. Phenolics and amino resins differ in the nature of the component which contains the reactive hydrogen atoms. Resins with phenols or aromatic compounds as base components will form phenolic polymers; if amines are used as base components amino polymers will result.

The other basic component is an aldehyde; formaldehyde is the preferred aldehyde for most resins. Higher aldehydes, such as acetaldehyde, may also be used. Ketones are possible components, but their reactivity is low.

7.4 Phenolic Resins

7.4.1 Basic Equations

Phenolic resins are the reaction products of phenol aromatic compounds with aldehyde, usually formaldehyde. The synthesis of the preliminary compound involves the following steps:

(a) Methylol phenols are formed in the first step. It is possible to introduce up to three methylol groups.

Phenol   +   Formaldehyde ⟶   Dimethylolphenol

The reactivities of ortho and para positions are nearly the same in an alkaline solution; in an acidic medium the para position is preferred. Methylol compounds are fairly stable during alkaline catalysis, but they react immediately during acidic catalysis, and the formaldehyde, already reactive, is activated even more. Proton addition onto the carbonyl oxygen transfers a full positive charge to the carbon atom of the formaldehyde:

$$CH_2O \quad + \quad H^+ \quad \longrightarrow \quad \overset{\overset{H}{|}}{\underset{\underset{H}{|}}{^+C}}-OH$$

During alkaline catalysis, on the other hand, the reactivity of the aldehyde is reduced. Simultaneously, phenol is activated by formation of phenolate, resulting in activation of the overall reaction.

Several different possibilities exist for the methylol compounds to react further:

(b) Condensation between methylol group and phenol:

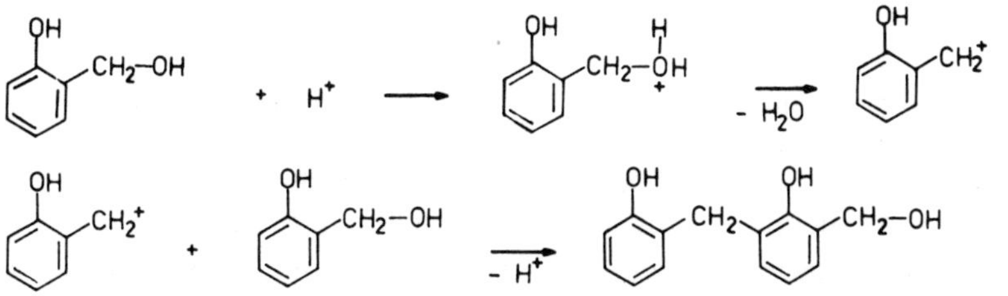

(c) In addition to the reaction mentioned in (b), self-condensation of methylol phenol is possible:

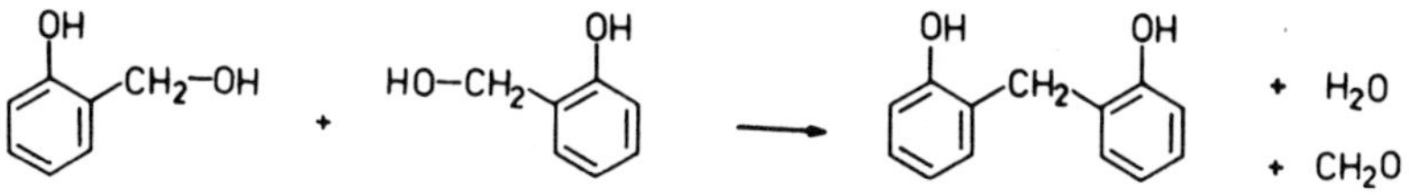

(d) An additional reaction is the formation of an ether:

(e) In contrast to components linked by methylene, the methyl ether groups are able to react further. The dimethylene ether group can eliminate one molecule of formaldehyde and form a methylene bond, especially during the thermosetting process. The formaldehyde generated by cleavage during this reaction is able to attack available reactive positions of the phenols.

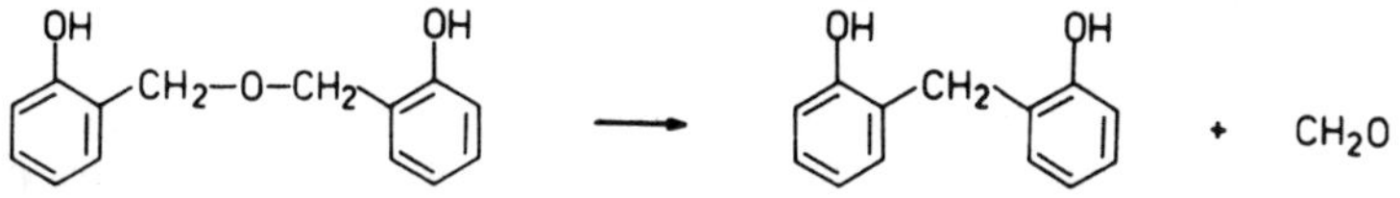

The conditions determine which of the reactions mentioned above will be the preferred one, depending on:

- The mole ratio of phenol to formaldehyde
- The pH value of the reaction solution
- The reaction temperature

During basic catalysis, methylol groups are more stable than during acidic catalysis. They convert slowly to dimethylene ether bonds. Neutralization of the resin mixture will stop the reaction to a large degree. Low temperatures and neutralization will slow down the reaction even more and make storage of the resin possible, if only for a limited time. During acidic catalysis, stable methylene groups are formed predominantly; the resins do not contain any methylol groups.

In addition to the reactions mentioned above, other reactions which involve "quinone methides" will be discussed:

quinone methide

It is possible for quinone methides to add onto phenol and methylol compounds. Furthermore, other reactions do occur, but only to a small extent. For instance, residual methylol groups are able to form aldehydes by cleavage of hydrogen; this may explain the release of hydrogen during the subsequent heat-treatment of cured parts [14].

Repeating successively steps (a) to (d) will lead to linear chain molecules. Since phenol has more than two reactive sites available and formaldehyde is bifunctional, the formation of highly cross-linked polymers is possible. The resulting polymers are completely insoluble and unmeltable.

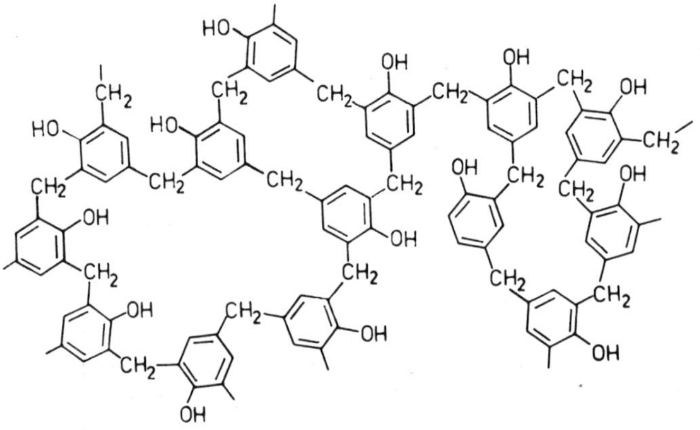

7.4.2   Raw Materials for Phenolics

The foremost raw materials of phenolic resins are low molecular weight phenolic compounds obtained as distillates during the coking process of coal. The phenols produced from tar are quite inexpensive but represent a mixture of components. Table 29 lists only those compounds available in coal tar which are phenols with only one benzene ring. Since the synthesis of phenolic resins results in only short, linear chains, cross-linking is necessary to obtain a product of high quality. Monofunctional components terminate the chains; bifunctional groups are able to form only linear chains; therefore, a high percentage of trifunctional components is necessary to form highly cross-linked polymers. Resorcinol, another important phenol compound, is also used as raw material.

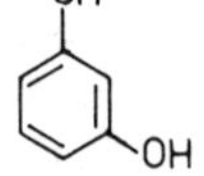

Resorcinol

Resorcinol contains two phenolic OH groups. The three hydrogen atoms in ortho and para positions may be exchanged easily, and resorcinol resins react quickly and at low temperatures. The high price of resorcinol is a drawback; resorcinol resins are used only for specialty polymers.

All phenolic compounds that have suitable physical properties and ortho- or para positioned hydrogen atoms can be used as base material. Undistillable coal tar, for instance, which is highly aromatic and contains phenolic OH groups, also belongs in this group.

Formaldehyde is used almost exclusively as the aldehyde component. It is a gas at room temperature and is used as an aqueous solution of 30–37% formaldehyde by weight. Oligomeric or polymeric formaldehydes, such as paraformaldehyde, trioxymethylene, or polyoxymethylene are meltable solids; they are unimportant to the synthesis of phenolic resins.

Common acids (mineral acids or carboxylic acids) or inexpensive alkaline compounds are used as catalysts for the synthesis and curing of precondensates. Oxalic acid is of significant value to the acid catalysis. Usually, ammonia or NaOH is used to achieve the necessary pH value during the alkaline catalysis.

7.4.3 Reactivity of the Starting Components

The high polarization of the formaldehyde molecule is the reason for its reactivity:

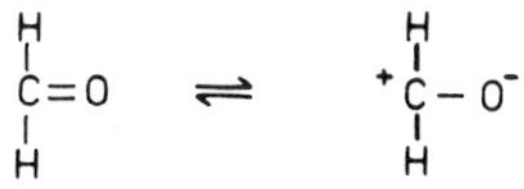

The positive charge on the carbon atom allows the formaldehyde molecule to enter into electrophilic reactions. Phenol is greatly susceptible to electrophilic reactions. Aromatic hydrogen atoms are activated by the phenolic OH group, which is an electron-attracting group and produces a dipole moment in the phenol molecule. This will increase the electron density at the carbon atoms in the benzene ring and facilitate dissociation of hydrogen atoms. In addition to the "inductive shift of the electrons", "mesomerism" exists which strongly activates the hydrogen atoms of benzene; it can be described as follows:

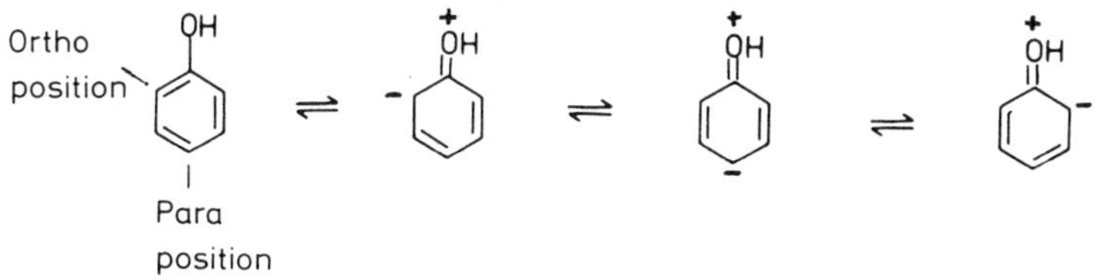

The "inductive" and "mesomer" effects point geometrically in the same direction and are additive. This observation leads to the prediction that phenol will react solely as a trifunctional compound with formaldehyde and reaction will take place only in ortho and para positions.

If the reaction occurs in an alkaline medium, phenol converts to the phenolate structure:

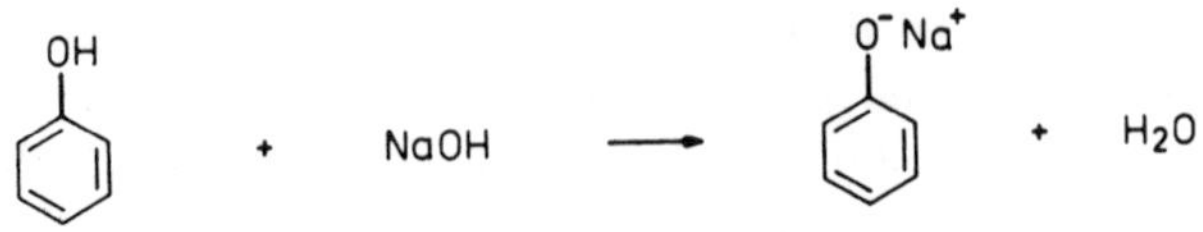

During synthesis or curing of phenolic plastics at high alkaline pH, a certain percentage of phenol will convert to phenolate. The negative charge of the phenolate ion increases the reactivity of the aromatic hydrogen atoms; this explains the catalytic reaction of alkali during the phenol-formaldehyde condensation reaction. Conversion to the phenolate structure does not change the reactive positions; this is evident from the mesomeric structures:

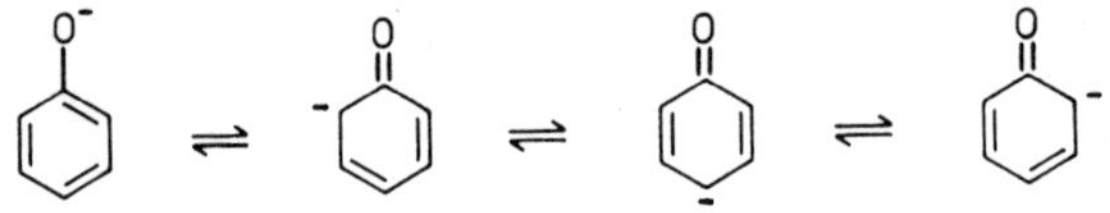

Cresols and xylenols are methyl-substituted phenols. A reactive position is blocked and the functionality of the phenolic component is less than 3 if the hydrogen atom in the ortho or para position is substituted by an alkyl group. The methyl group (generally speaking, all alkyl groups) also, exhibits an "inductive" effect which pushes electrons into the benzene ring. If the OH and methyl groups are in the meta position, the "inductive" effect of the methyl group will be additive to the "inductive" and "mesomer" effects of the hydroxyl group, and the reactivity of the compound will be increased.

Table 29   Relative reactivity and relative rate of resinification of important phenols with formaldehyde, compared to the reactivity and rate of reaction of phenol equal 1 (according to [11, 15, 16]

Compound	Name	Relative reactivity	Relative rate of resinification
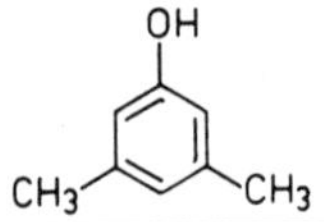	3,5 Xylenol	7,75	8,5
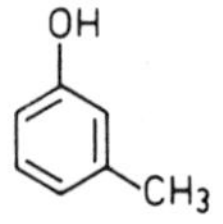	m-Cresol	2,88	3,4
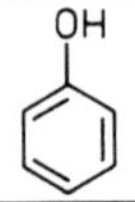	Phenol	1,0	1,0
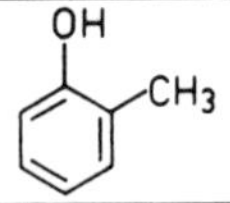	o-Cresol	0,76	0,4
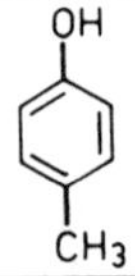	p-Cresol	0,35	0,6
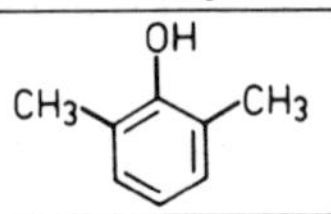	2,6-Xylenol	0,16	

On the other hand, the reactivity of the compound decreases if a methyl group is in an ortho or para position, since methyl and OH groups cause opposite effects. This theory conforms well to the values found in practice (see Table 29).

Reactive hydrogen atoms are replaced by other groups if alkyl substituents are in ortho and/or para positions. The functionality of the components decreases with decreasing number of exchangeable reactive hydrogen atoms. Therefore, ortho- and para-cresol become bifunctional compounds; 2,4-xylenol and 2,6-xylenol have only one reactive site remaining (monofunctional), and 2,4,6-trimethyl phenol does not react at all with formaldehyde.

A resin with a specific reactivity can be prepared by mixing different phenolic raw materials. Also, the amount used of mono- and bifunctional compounds will affect the strength of the end product. The polymer processor has to consider the different functional groups and the reactivities of the various resin components; he has to be aware that indiscriminate mixing and adding of resins will have an effect on the mechanism of the curing reaction and on the properties of the final product.

7.4.4   Self-Curing Resins and Resins with a Long Shelf Life

The reaction between phenol and formaldehyde in a molar ratio of 1:1 in an acidic reaction medium yields products in which the phenol is mostly cross-linked by methylene groups. The initially formed methylol groups react further until none of the reactive groups are left in the resin. The result is a product with an unlimited shelf life. Up to 13 phenol components can be linked by methylene groups and the resulting molecular weights are between 1200 and 1400 [17]. These products are called novolaks.

They are synthesized in an aqueous solution at 100 °C. After the reaction is completed, the resin will form a separate resin layer at 80 °C. The pH value of the resin, which still contains some water, is adjusted to 6.0–6.5. The remaining water is removed by distillation, and the resulting resin has a relatively long shelf life.

Resols are synthesized by reacting a mixture of phenol and formaldehyde in an alkaline solution. The ratio of phenol to formaldehyde is 1: >1. For the formation of the molding mass, a molar excess of formaldehyde of up to 50% is used and condensation takes place above 70 °C. The resins formed are mostly linear and contain a certain percentage of methylol groups. As a result of the higher temperature, phenol resins are cross-linked mostly by methylene groups. It is advantageous for the polymer to contain a minimum amount of ether bonds, since the dimethyl ether groups are responsible for the liberation of formaldehyde during thermo setting.

Resols synthesized in aqueous solutions will be neutralized and, just as in the synthesis of novolaks, water is driven off. Resols melt easily and are soluble in a multitude of organic solvents; they are thermoplastic A-stage resins. In contrast to the stable novolaks, resols are reactive and have a limited shelf life. During heating, the reaction is carried further and resols are converted to resitols, the B stage. Resitols will swell in organic solvents but do not melt; they are in a highly viscous, rubberlike state. Increasing the temperature increases the cross-linking, and resitols convert to resites. Resites neither swell in organic solvents nor are they fusible. Resites represent the so-called C stage.

Novolaks or resols are processed by the polymer processor either as pure resins or as resins containing various fillers. The products show different shelf life properties:

Novolak: Unlimited shelf life
Resol: Reactive; limited shelf life

Also, the ease with which the product can be processed is different for these polymers. Resols are reactive intermediates, stabilized by lowered temperatures and neutralization; in the polymer processing plant, the condensation reaction can be continued by adding a catalyst and raising the temperatures. Novolaks, on the other hand, have to be mixed with a hardener such as hexamethylenetetramine ("hexa") and cross-linked by heating.

Resols may be cured at lower temperatures if a catalyst is present. Neutral resins can be cured by heating. During the curing process, the remaining methylol groups react either with each other or with an aromatic cyclic structure. Water liberated by this process is detrimental to the quality of the product; as is the corrosive effect of the acid or base used as a catalyst.

The curing process of novolaks, on the other hand, depends on a curing agent and involves a different reaction mechanism. It is not a continuation of an interrupted condensation reaction; but here too, small amounts of volatile cleavage products are liberated.

7.4.5   Curing of Novolaks

Novolaks are phenols crosslinked mostly by methylene groups and cannot be cured any further by reactive groups. Therefore, it is necessary to add curing agents which allow the novolaks to react further. Hexamethylenetetramine is the most widely used curing agent. Hexamethylenetetramine is formed by reacting formaldehyde with ammonia and is able to form the cleavage products, formaldehyde and ammonia, by reverse synthesis.

$$6\ CH_2O\ +\ 4\ NH_3\ \rightleftharpoons\ \begin{array}{c}\text{hexamethylenetetramine}\end{array}\ +\ 6\ H_2O$$

In the presence of novolaks and at temperatures of 150–180 °C, hexamethylenetetramine reacts with the available reactive groups of the (phenolic) cyclic structure by forming

dimethylene amino bridges $- CH_2 - NH - CH_2 -$

trimethylene amino bridges $- CH_2 - N\big\langle \begin{array}{c} CH_2 - \\ CH_2 - \end{array}$

A small part of the nitrogen present in the curing agent will be released as ammonia, with even the most favorable curing conditions. In the presence of a large excess of phenolic compounds, the nitrogen-containing cross-linkages react further and form methylene bridges, thus releasing additional ammonia.

The "hexa" curing has the advantage of being water-free, and only small quantities of ammonia are released if the right ratio of components is used. But even small amounts of ammonia will have a corrosive effect on tools and metallic articles which are in contact with the freshly cured article.

At higher temperatures, cleavage of hydrogen from the dimethyleneamine cross-linking molecule results in formation of a double bond which is in conjugation with the aromatic ring structure and is responsible for the yellow color of articles cured with "hexa."

$$-CH_2-NH-CH_2 \longrightarrow -CH=N-CH_2- + H_2$$

Curing with formaldehyde is possible, in principle. The boiling point of formaldehyde is $-21\,°C$; at room temperature it is a gas and is difficult to handle; usage is limited to aqueous solutions. Diffusion of formaldehyde into the polymer may result in curing; but this is not a customary process for the manufacture of phenolic resins. Pulverized paraldehyde, formaldehyde in oligomer form, can easily, and without the use of water, be mixed with the resin. But the curing reaction proceeds very slowly, and a catalyst is necessary; only the very reactive resorcinol resins can easily be cured by paraldehyde. Additional compounds used as cross-linking agents are components high in methylol groups, such as trimethylol phenol, dimethylol urea, or hexamethylol melamine. The reactions of the curing process are similar to the reactions during resin synthesis. Compared to the "hexa curing," these processes have the disadvantage of liberating water.

7.4.6   Processing of Phenol Formaldehyde Resins

Phenol formaldehyde resins processed in the polymer processing plant can be classified as either casting resins or molding compounds. Usually, casting resins are resols which are liquid at lower temperatures and have an acceptable shelf life. To facilitate curing, acid is added to the neutral resin in order to lower the pH value to the point at which further condensation occurs and results in total curing and cross-linking of the polymer. Adding acid produces the same reaction medium as exists during the synthesis of resols.

For more economical curing times, higher temperatures and the use of curing agents are necessary. The pH value of the resin is determined by the acid that is used as the curing

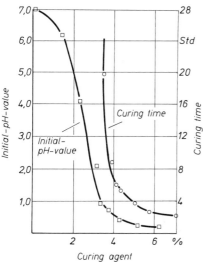

Figure 19
Curing times and initial pH-value of a phenol-resol resin as a function of the concentration of the curing agent [18].

agent; the greater the amount of curing agent added, the lower the pH value will be. The rate of condensation increases with decreasing pH value (Figure 19) and increasing temperature (Figure 20). A sufficiently fast curing is possible only below a certain pH value. If only small quantities of catalyst are used, then the self-setting becomes more dependent on temperature (Figure 20). The strictly thermal curing of a resin at pH 7 requires high temperatures, and the curing rate depends strongly on the temperature (Figure 21).

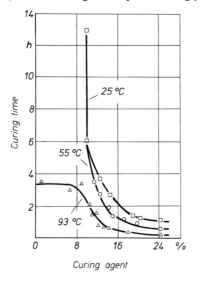

Figure 20
Curing time of a phenol resol at various temperatures as a function of the concentration of the curing agent [18].

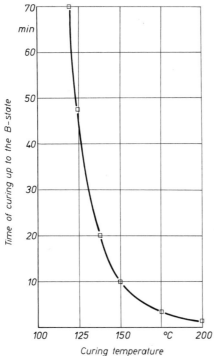

Figure 21
Curing rate of a phenolresol as a function of the curing temperature [11].

Water formed during the condensation reaction will lead to voids in the molded article, especially in resins without fillers, if contact molding and high temperatures are employed. Catalytic curing using acid and low temperatures will result in polymers which are almost void-free, but it has the disadvantage of employing corrosive acids which affect the quality of the end product.

Resols synthesized from a mixture of phenols, cresols, and/or xylenols have a different curing rate than those made from phenols only. The curing rate depends on the reactivity of the raw material. Resols synthesized by using larger amounts of resorcinol cure quickly without the addition of acid and at moderately high temperatures. The corrosive auxiliary agent is not necessary. These resins are of very high quality and are expensive.

Novolaks that are solid at room temperature and easily fusible in a hot mold are preferred as molding materials. They are used almost exclusively with large quantities of filler material.

The curing of novolaks by hexamethylenetetramine requires a certain minimum amount of the curing agent to achieve satisfactory results. The required quantity must be determined experimentally by finding the optimum quality of the end products. The curing rates of novolaks depends little on the concentration of the curing agent as long as the necessary minimum is exceeded (Figure 22). Curing temperatures lie between 140 and 180 °C.

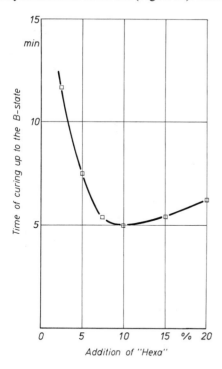

Figure 22
Curing rate of a phenol novolak as a function of quantity of hexamethylenetetramine added [11].

Resols too can be used as molding compounds and cured at the same temperatures, but a high molding pressure is necessary to keep the water, formed as a by-product, in solution.

7.4.7   Heat of Reaction and Shrinkage

The conversion of phenol and formaldehyde to a macromolecular compound is accompanied by the release of a large amount of heat. With the exception of the resins that are synthesized from phenolics used for the production of foam, the resins used in the polymer processing plant are prereacted to the extent that only a very few molar conversions take place during curing. The heat of reaction liberated during the curing of cast resins and molding compounds is of little consequence in the fabrication of molded articles.

As a result of the small amount of molar conversion, the reaction related shrinkage is insignificant during the curing of cast resins and molding compounds. It has to be taken into consideration that water formed as a by-product during the curing of resol remains in the end product. The post-curing shrinkage of phenol formaldehyde resins cured at temperatures above 100 °C and for long periods of time can be so high that it is of operational importance. The polymer shows a weight loss (Figure 23), which is probably the result of evaporation of the entrapped water. The degree of "post-curing shrinkage" is approximately

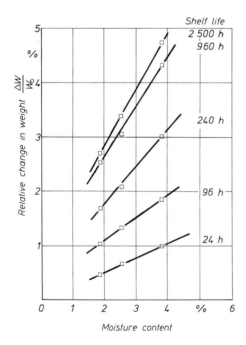

Figure 23
Relative change in weight of a phenol-formaldehyde molding compound after tempering at 140 °C as a function of shelf life and moisture content [14].

Table 30 Composition of all volatile compounds during tempering of phenolic resin molding compounds (268 hr at 130 °C; 4.13% weight loss) [14].

Substance	% by mole	% by weight
H_2O	98,82	98,10
H_2	0,08	0,01
N_2	0,74	1,10
CO_2	0,17	0,45
CH_3OH	0,09	0,15
O_2	0,10	0,19

proportional to the tempering time and the percentage of water in the molding compound at the time of removal from the mold [14]. The analysis of compounds liberated during tempering shows that almost no other substances are released except water and the entrapped air.

7.4.8 Phenol Furfural Resins

Phenol furfural resins have importance as polymers which are free-flowing, and moldable by either casting or injection molding. They are synthesized by condensation of phenol, furfural, and formaldehyde; two different aldehydes are involved in this reaction.

Furfural

In the first stage phenol is reacted with furfural. The condensation uses alkali and is slower than the phenol formaldehyde condensation. In addition to the condensation reaction, polymerization of furfural can lead to resinification. This is avoided by using an excess of phenol, which causes A-stage resins (resols) to be formed; in the following stage they react with formaldehyde and carry the condensation reaction further.

The curing of phenol furfural resins takes longer than the curing of phenol formaldehyde resins.

7.4.9 Phenolic Resins – Foams

The manufacture of foam plastics uses resols synthesized by alkaline condensation; they are used in the liquid state and may contain a small percentage of water. The resols are mixed with the foaming agents. In principle, decomposition agents as well as evaporation agents may be used as foaming agents. Mostly low boiling volatile foaming agents such as low molecular weight paraffins or halogenated low molecular weight hydrocarbons are used.

The resin can be cured either by heating or by catalysis in cold molds. For thermosetting, the resin is mixed with the catalyst (acid) and the foaming agent and then filled into preheated molds. The heat transfer from the hot walls of the molds to the resin accelerates the curing. External heat can cause evaporation or decomposition of the foaming agent; therefore, foaming is not directly related to the reaction of the components or the heat liberated by it. After the foam formation is completed quick curing of the resin is necessary in order to stabilize the foam.

For the cold curing, very reactive, slightly prereacted resols are needed. Until the end of the foaming process is reached, the reacting mass is not allowed to become rigid. Therefore, cold setting requires carefully selected ratios of resin, curing agent, and foaming agent. The higher percentage of shrinkage is unimportant because most of the shrinkage takes place before the polymer becomes rigid and allows stress to develop.

Bibliography to Sections 7.2–7.4

[1] *N. N.:* Kunststoffe 69 (1979) 535.

[2] *Hultzsch, K.:* »Chemie der Phenolharze«, Springer Verlag, Berlin, 1950.

[3] *Martin, R. W.:* »The Chemistry of Phenolic Resins«, Wiley & Sons, Inc., New York, 1956.

[4] *Wegler, R., H. Herlinger:* »Polyadditions- und Polykondensationsprodukte von Carbonylverbindungen mit Phenolen«, in Houben-Weyl: Methoden der organischen Chemie: 4. Aufl. Edited by E. Müller, Bd. XIV/2: Makromolekulare Stoffe, Thieme-Verlag, Stuttgart, 1963, 193.

[5] *Carswell, T. S.:* »Phenoplasts, Their Structure, Properties and Chemical Technology«, Interscience Publ. Inc., New York, 1947.

[6] *Frisch, K. C.:* »Phenolic Resins and Plastics«, in Encyclopedia of Chemical Technology, Bd. 10, edited by R. E. Kirk and D. F. Othmer, Interscience Encyclopedia Inc., New York, 1953.

[7] *Gould, D. F.:* »High Temperature Plastics«, edited by W. Brenner, D. Lum and M. W. Riley, Reinhold Publ. Corp., New York, 1962.

[8] *Gould, D. F.:* »Phenolic Resins«, Reinhold Publ. Corp., New York, 1959.

[9] *Greth, A., F. Lemmer:* »Phenolharze« in Ullmanns Encyklopädie der techn. Chemie, 13. Bd., Urban & Schwarzenberg, München-Berlin, 1962.

[10] *Knop, A., W. Scheib:* »Chemistry and Application of Phenolic Resins«, Springer-Verlag, Berlin-Heidelberg-New York, 1979.

[11] *Vieweg, R., E. Becker:* »Duroplaste«, Kunststoff-Handbuch, Bd. X, Carl Hanser Verlag, München, 1968.

[12] *Wegler, R., H. Herlinger:* In: Houben-Weyl: Methoden der organischen Chemie, Bd. XIV/2, Makromolekulare Stoffe, Thieme Verlag, Stuttgart.

[13] *N. N.:* Mod. Plast. Intern. 11 (1981) 33.

[14] *Behmer, A.:* Dissertation at the Institut für Kunststoffverarbeitung, RWTH Aachen, 1975.

[15] *Sprung, M. M.:* J. Amer. Chem. Soc. 63 (1941) 334.

[16] *Freeman, J. H., C. W. Lewis:* J. Amer. Chem. Soc. 76 (1954) 2080.

[17] *Müller, H. F., I. Müller:* Kunststoffe 38 (1948) 221.

[18] *Little, G. E., K. W. Pepper:* Brit. Plastics 19 (1947) 430.

7.5 Amino Resins

7.5.1 Basic Chemical Equations

The classification "amino resins" includes all molding compounds which are synthesized by the polycondensation of amines and aldehydes. As with the phenolic resins, the initial reaction is an addition, a nucleophilic attack of the aldehyde on a compound containing reactive hydrogen atoms which yields methylol compounds, and the second step is a polycondensation, during which the methylol compounds polymerize to high molecular weight amino resins by elimination of water [1–4].

During synthesis of phenolic resins the formaldehyde reacts with aromatic hydrocarbons and the reactive hydrogen atom is an activated hydrogen atom of an aromatic structure. On the other hand, formaldehyde reacts with amino, amide, or imide groups during formation of amino resins. The general chemical equation, described in Section 7.3, must be expanded as follows in order to include the synthesis of amino resins:

$$H_2N-\bigcirc-NH_2 \quad + \quad CH_2O \quad \longrightarrow \quad H_2N-\bigcirc-NH-CH_2-OH$$

$$H_2N-\bigcirc-NH-CH_2-OH \quad + \quad H_2N-\bigcirc-NH_2 \quad \longrightarrow$$

$$H_2N-\bigcirc-NH-CH_2-NH-\bigcirc-NH_2 \quad + \quad H_2O$$

Macromolecules are formed by continued reaction between formaldehyde and amino groups as well as by further condensation of methylol groups. A strictly linear condensation would produce a polymer of little commercial use, but at favorable conditions cross-linked polymer is produced, since the amines used for the synthesis of the amino resin usually contain more than one NH_2 group and both their hydrogen atoms can be replaced by methylol groups; however, this is not always possible.

7.5.2 Raw Materials for Amino Resins

Urea, melamine, and formaldehyde are the most important raw materials for the production of casting resins and molding compounds made from amino resins. The synthesis and curing of urea and melamine resins show similar chemical reactions. As in the production of phenolic resins, formaldehyde is the most reactive and important aldehyde. Since it is also very inexpensive, the synthesis of amino resins uses formaldehyde almost exclusively.

Thiourea, alone or as a mixture with urea or melamine, is a suitable component for the synthesis of amino resins which can be molded and cured without a curing catalyst. These resins are more stable toward hydrolysis than articles made from pure urea formaldehyde resin. But since these products release formaldehyde on contact with water, their application is limited [5]. Dicyandiamide and aniline resin molding compounds are also of little importance. Aniline is an amine, but it reacts as a phenol when in contact with formaldehyde. According to the following resonance structures, formaldehyde will react with aniline in the same way as with phenol, by formation of methylol groups in positions ortho and para to the amino group; the methylol groups will then undergo further condensation reaction.

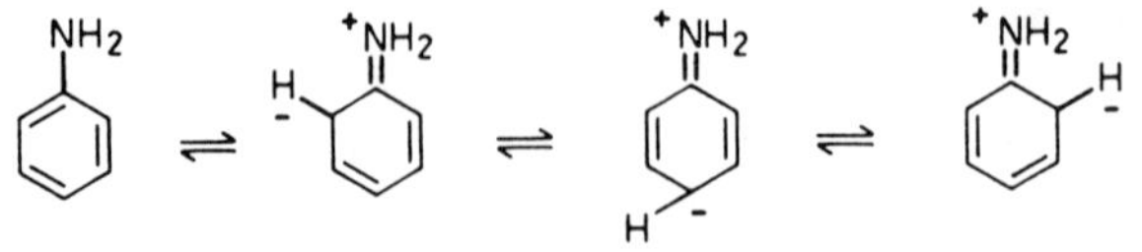

Table 31   Raw materials needed for the synthesis of amino resins.

Formula	Name	Melting point [°C]
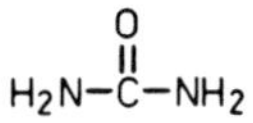	Urea	132,7
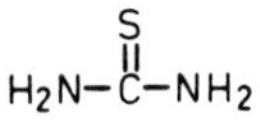	Thiourea	180
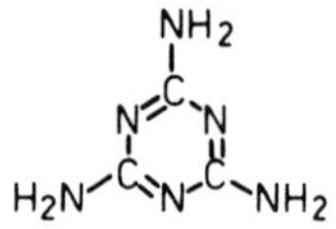	Melamine	ca. 250
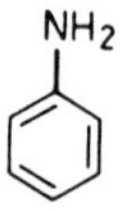	Aniline	−6,0
CH_2O	Formaldehyde	Sdp. −21 °C
CH_3-CHO	Acetaldehyde	Sdp. +20,2 °C
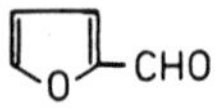	Furfural	Sdp. +161,6 °C

7.5.3   Reactions during Synthesis of Amino Resins

Urea reacts with a maximum of 3 moles of formaldehyde [6] and forms mono-, di-, and/or trimethylol urea:

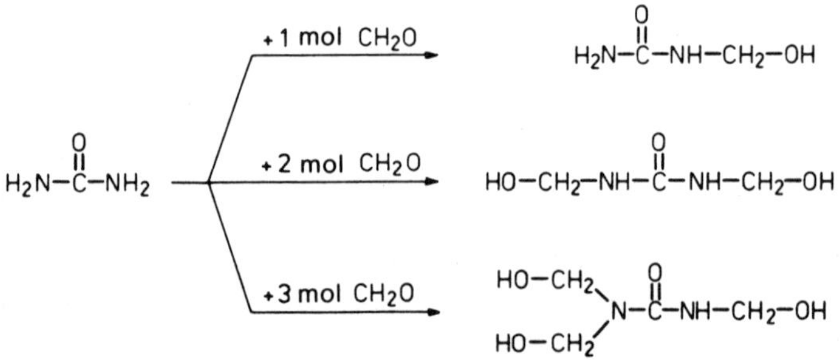

Melamine reacts with a maximum of 6 moles of formaldehyde (if enough formaldehyde is available) and forms hexamethylol melamine [7]:

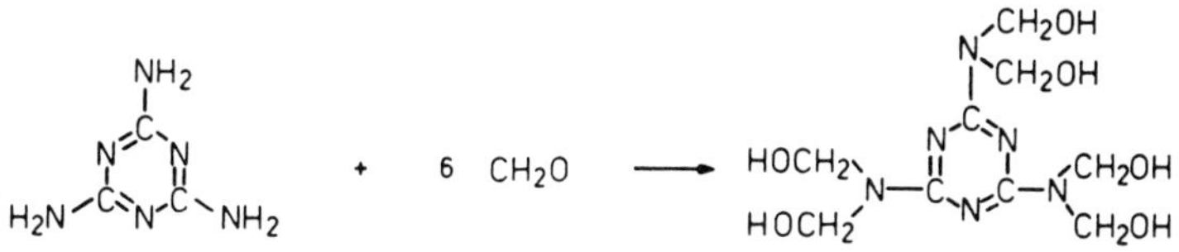

The first step in the synthesis of amino resins is the formation of methylol compounds. This is strictly an addition reaction; it is accelerated catalytically by both acids and bases. The addition of acids causes protonation of the formaldehyde and increases its reactivity toward amines. Likewise, in an acidic medium urea will convert to a protonized form which shows lower reactivity toward formaldehyde than regular urea does. Overall, the increase in the reactivity of formaldehyde is predominant; hence, acids increase the rate of reaction of formaldehyde with urea.

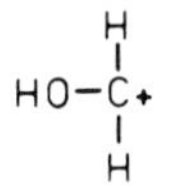

$$HO-\overset{\overset{H}{|}}{\underset{\underset{H}{|}}{C}}+ \qquad \text{Protonized formaldehyde}$$

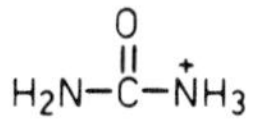

$$H_2N-\overset{\overset{O}{\|}}{C}-\overset{+}{N}H_3 \qquad \text{Protonized urea}$$

Alkalis increase the reactivity of urea by removing a proton:

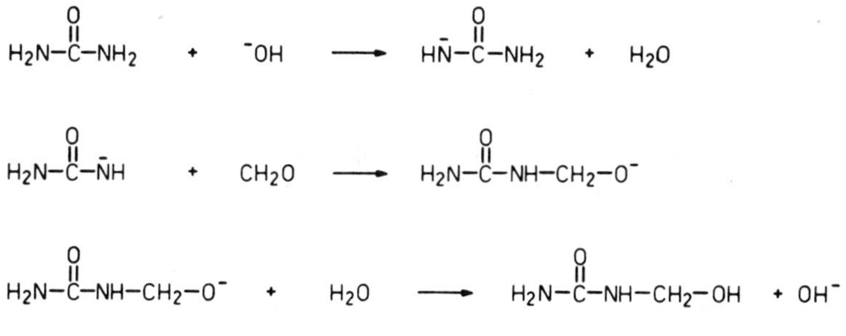

$$H_2N-\overset{\overset{O}{\|}}{C}-NH_2 \quad + \quad {}^-OH \quad \longrightarrow \quad H\overset{-}{N}-\overset{\overset{O}{\|}}{C}-NH_2 \quad + \quad H_2O$$

$$H_2N-\overset{\overset{O}{\|}}{C}-\overset{-}{N}H \quad + \quad CH_2O \quad \longrightarrow \quad H_2N-\overset{\overset{O}{\|}}{C}-NH-CH_2-O^-$$

$$H_2N-\overset{\overset{O}{\|}}{C}-NH-CH_2-O^- \quad + \quad H_2O \quad \longrightarrow \quad H_2N-\overset{\overset{O}{\|}}{C}-NH-CH_2-OH \quad + \quad OH^-$$

The addition of formaldehyde to urea is an equilibrium reaction [8].

$$\frac{[\text{Formaldehyde}]\cdot[\text{Urea}]}{[\text{Methylol urea}]} = K$$

[ ] = concentration
K = equilibrium rate constant

Similar equilibrium reactions can be formulated for other methylol compounds. The equilibrium is very far to the side of the methylol compound; nevertheless, release of formaldehyde is possible.

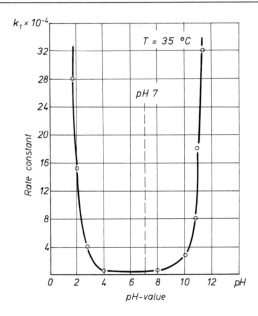

Figure 24
Influence of pH value on the rate
constant of formaldehyde urea
reaction at 25 °C [9].

Methylol compounds formed by adding formaldehyde onto amine compounds are
weak bases and are able to form a methylene cation:

$$HO - CH_2 - NH - \underset{\underset{O}{\overset{\|}{C}}}{} - NH - CH_2 - OH \quad \xrightarrow[- H_2O]{+ H^+} \quad$$

$$HO - CH_2 - NH - \underset{\underset{O}{\overset{\|}{C}}}{} - NH - \overset{+}{C}H_2$$

$$\updownarrow$$

$$HO - CH_2 - NH - \underset{\underset{O}{\overset{\|}{C}}}{} - \overset{+}{N}H = CH_2$$

Several nucleophilic compounds, such as urea, amides, or alcohols, can be added onto
the cation formed by the reaction with acid. The resulting growth of the molecule will even-
tually produce macromolecules by polycondensation.

$$HO-CH_2-NH-\underset{\underset{O}{\overset{\|}{C}}}{}-NH-CH_2-OH$$

$$+ H^+$$
$$- H_2O \downarrow$$

$$HO-CH_2-NH-\underset{\underset{O}{\overset{\|}{C}}}{}-NH-\overset{+}{C}H_2$$

+ $H_2N-\underset{\underset{O}{\overset{\|}{C}}}{}-NH_2$ (Urea) ⟶ $HO-CH_2-NH-\underset{\underset{O}{\overset{\|}{C}}}{}-NH-CH_2-NH-\underset{\underset{O}{\overset{\|}{C}}}{}-NH_2$ + H^+

+ $\underset{\underset{CH_2-OH}{}}{NH-\underset{\underset{O}{\overset{\|}{C}}}{}-NH-CH_2OH}$ (Dimethylol urea) ⟶ $HO-CH_2-NH-\underset{\underset{O}{\overset{\|}{C}}}{}-NH-CH_2-\underset{\underset{CH_2-OH}{}}{N}-\underset{\underset{O}{\overset{\|}{C}}}{}-NH-CH_2-OH$ + H^+

+ $HO-R$ (Alcohol. Methylol group) ⟶ $HO-CH_2-NH-\underset{\underset{O}{\overset{\|}{C}}}{}-NH-CH_2-O-R$ + H^+

The reaction between the methylene cation and alcohols is important not only for the
condensation of methylol compounds which will result in macromolecules and cross-linkage but also for possible reactions with the hydroxyl groups of cellulose (wood flour is used
as filler for molding compounds) and with modifiers containing hydroxyl groups. The ability to react with various fillers is of particular practical importance, since it is the basis for
excellent adhesion between filler and matrix.

All the reactions mentioned above are also applicable to melamine-formaldehyde reactions. Melamine is more reactive than urea, since the aromatic triazine molecule increases the reactivity of the amino and methylol groups.

Condensation of methylol compounds results in compounds with high molecular weight. Heat alone will start this condensation of the thiourea and melamine compounds. Alkaline or acidic catalysts are needed for further condensation of methylol urea resins. In an alkaline medium (pH 8), the polycondensation of methylol urea resins will result in methylene-ether linkages [10]; condensation in a medium of pH 5 will form methylene groups [11].

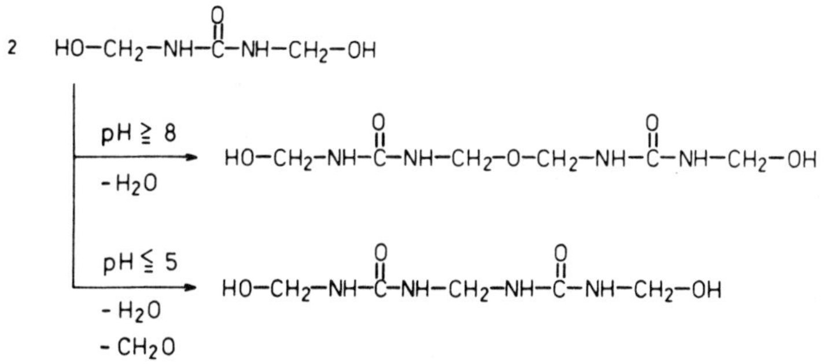

$$2 \quad HO-CH_2-NH-\overset{\overset{\displaystyle O}{\|}}{C}-NH-CH_2-OH$$

$$\xrightarrow[-H_2O]{pH \geq 8} \quad HO-CH_2-NH-\overset{\overset{\displaystyle O}{\|}}{C}-NH-CH_2-O-CH_2-NH-\overset{\overset{\displaystyle O}{\|}}{C}-NH-CH_2-OH$$

$$\xrightarrow[\substack{-H_2O \\ -CH_2O}]{pH \leq 5} \quad HO-CH_2-NH-\overset{\overset{\displaystyle O}{\|}}{C}-NH-CH_2-NH-\overset{\overset{\displaystyle O}{\|}}{C}-NH-CH_2-OH$$

Further, linear condensation in an alkaline medium may yield the following product:

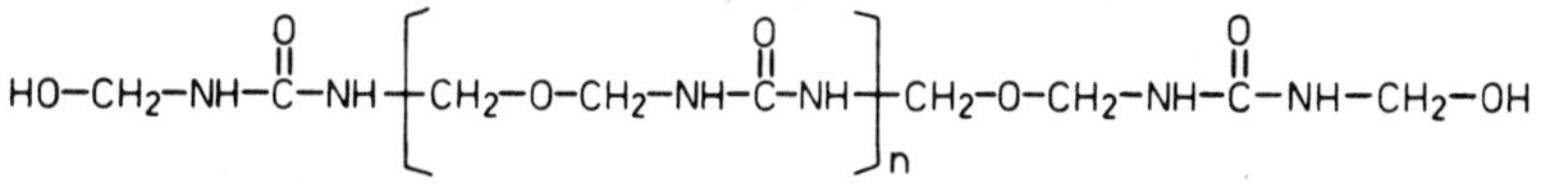

$$HO-CH_2-NH-\overset{\overset{\displaystyle O}{\|}}{C}-NH\left[CH_2-O-CH_2-NH-\overset{\overset{\displaystyle O}{\|}}{C}-NH\right]_n CH_2-O-CH_2-NH-\overset{\overset{\displaystyle O}{\|}}{C}-NH-CH_2-OH$$

This product will be formed predominantly in an acidic medium:

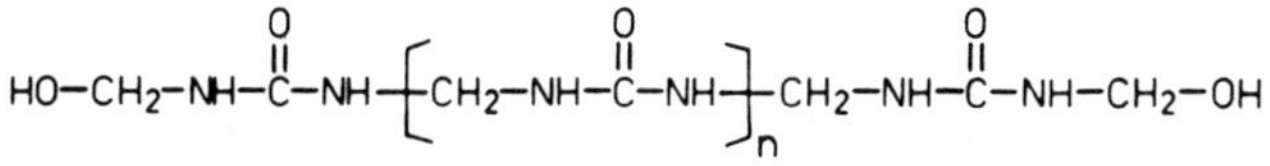

$$HO-CH_2-NH-\overset{\overset{\displaystyle O}{\|}}{C}-NH\left[CH_2-NH-\overset{\overset{\displaystyle O}{\|}}{C}-NH\right]_n CH_2-NH-\overset{\overset{\displaystyle O}{\|}}{C}-NH-CH_2-OH$$

The polyfunctionality of urea together with the reactivity of the acidic amide group will cause cross-linking between the propagating molecules as follows:

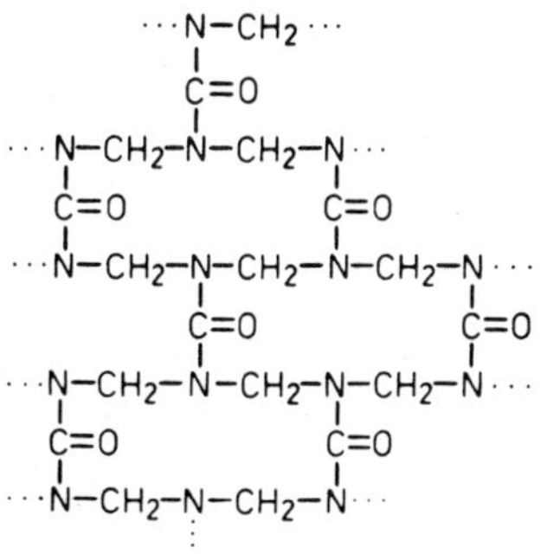

```
                   ···N−CH2 ···
                      |
                      C=O
                      |
         ···N−CH2−N−CH2−N ···
            |           |
            C=O         C=O
            |           |
         ···N−CH2−N−CH2−N−CH2−N ···
               |           |
               C=O         C=O
               |           |
         ···N−CH2−N−CH2−N−CH2−N ···
            |           |
            C=O         C=O
            |           |
         ···N−CH2−N−CH2−N ···
                 ⋮
```

7.5.4   Manufacture of the Resin

In principle, the manufacture is the same for urea formaldehyde resins and melamine formaldehyde resins. The resin condensation takes place in agitated vessels at temperatures of 20–100 °C and pH values of 3.5–8.5. Urea is dissolved in a 40% aqueous solution of formaldehyde; reaction between the two components occurs immediately. Melamine is only slightly soluble in an aqueous formaldehyde solution. The newly formed methylol compounds, however, are soluble.

Several manufacturing processes are available. They differ in temperature, in pH value, and in the quantity of the amine component which is added to the solution of formaldehyde. Different conditions yield resins with different properties; resins will differ in molecular weight, percentage of methylol groups present, and arrangement of methylol groups. Cooling and neutralization of the reaction mixture terminate the condensation reaction.

Most amino resins are processed in a mixture with fillers. Cellulose is the most widely used filler. Various methods are available for incorporating the filler in the resin. The "how" and "when" of mixing resin and filler will have an effect on the properties of the articles made from molding compounds. Three different methods are available for incorporation of fillers into the amino resins; the same is true for the manufacture of filled phenolic resins.

(a) Condensation (reaction) in the presence of the filler:

Condensation of the resin is carried out in a reactor vessel to a certain point where the percentage of products of low viscosity and low molecular weight is still quite high. The filler is added to this solution; the low molecular weight polymer compounds are able to flow into the smallest cracks of the filler. The condensation proceeds until the required conversion is reached; then water is removed from the resin-filler mixture by a vacuum kneader or heated rolls.

(b) Aqueous impregnation:

At the intended conversion stage, the aqueous resin solution is mixed with the filler. The soluble precondensate is able to wet the filler quite well. Reactive fillers such as cellulose are able to graft the resin molecules to the filler; the result is an excellent composite of resin and matrix in the cured end product.

The filler is mixed with a 40–50% aqueous resin solution in kneaders. After adequate mixing, the mixture is carefully dried; premature curing must be avoided. Also, drying can take place in a vacuum kneader or disk dryer. Afterwards, the material is pulverized, and the resulting powder is compacted and granulated to the molding compound.

(c) Dry impregnation:

The separately synthesized and dried intermediate is mixed with fillers and other additives in kneaders or internal mixers. The resulting crumbs are plasticized in rolls where the filler is homogeneously distributed. The cooled, rough sheets are crushed and ground in ball mills.

Resins combined with the filler during synthesis or mixed with it in an aqueous solution will result in articles that show better adhesion between matrix and filler than molding compounds prepared by dry impregnation. This is as true for fillers (stone powder) which do not react with the resin (stone flour) and form a matrix-filler bond based on physical or mechanical effects as for reactive fillers such as cellulose. The strength of the finished article increases with increasing strength of the interface of the resin-matrix filler bond.

7.5.5 Acid Catalysts Used in Resin Curing

An increase in temperature will accelerate the condensation of thiourea and melamine resins sufficiently, and under neutral conditions the resins can be cured by heating alone.

Heat alone is not sufficient to cure urea resins; acidification to a pH of 6.5 to 5.0 is necessary. It is usually impossible to reach the necessary pH value just by mixing acid and the stable, neutral resin shortly before the curing process. An excellent method is to use "latent" curing catalysts. They are added to the resin at the end of the synthesis and release the acid at molding temperature.

The variety of these "masked acids" is great. They show different activities, and specific activators with optimum activity are used in various ways to manufacture the resins. Ammonium chloride is a typical acid-releasing compound. Thermal cleavage yields ammonia and hydrochloric acid:

$$NH_4Cl \xrightarrow{\text{Heat}} NH_3 \quad + \quad HCl$$

Ammonium chloride	Ammonia	Hydrochloric acid

Metal salts made from weak bases and strong acids (such as $ZnCl_2$) are decomposed by heat and water, yielding a hydrated metal oxide and an acid. Another method to produce acid is the use of oxidizing agents such as sodium nitrate. According to the law of mass action, the decomposition of methylol compounds will generate small quantities of formaldehyde during the curing of amino resins. Heat and strong oxidation agents quickly oxidize formaldehyde to formic acid, which is the actual catalyst.

$$NaNO_3 \quad + \quad \underset{H}{\overset{H}{>}}C=O \longrightarrow NaNO_2 \quad + \quad H-C\underset{OH}{\overset{O}{<}}$$

Sodium nitrate	Formaldehyde	Sodium nitrite	Formic acid

Since these resins are reactive intermediates, their shelf life is limited. Even at room temperature, curing by polycondensation progresses slowly and leads to gelation of the polymer and a decrease in flow (figure 25, see next page). Storage at the lowest possible temperatures will increase their shelf life.

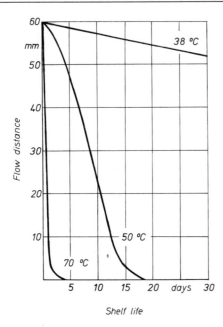

Figure 25
Shelf life of amino resin molding compounds at 38, 50 and 70 °C [12].

7.5.6 Resin Cure

The curing of amino resins is a continuation of the condensation reaction started by the raw material producer. The oligomer and polymer methylol compounds produced by the raw material producer undergo, during curing, the same reaction as during the synthesis of the resin. Polycondensation of compounds with at least three methylol groups will lead to cross-linked macromolecules. Besides the dimethylene ether cross-linkages formed by eliminating water, methylene linkages are formed by the elimination of water and formaldehyde. Immediately, the formaldehyde may react with NH_2- or NH-groups and form methylol groups; this will produce additional cross-linkage. Heat is necessary for the curing process; urea resins are cured at 130–160 °C and melamine resins at 140–170 °C.

The conditions of the curing process depend on the respective resin. The curing times required can be determined only from physical or mechanical properties. Usually, condensation is not totally finished at removal (from the mold). Dimensional stability and resistance to boiling water are used as criteria for curing times and depend on the temperature (see Figure 26).

Amino resins have to be cured under pressure (25–50 N/cm²). This pressure is necessary in order to get smooth surfaces and to keep the water formed during condensation sufficiently finely dispersed in the article.

The molding compounds contain lubricants and plasticizers to improve their flow properties. In practice, the water contained in the resin or produced during curing presents great problems. It increases the resin flow, but it is responsible for molding defects and post-curing shrinkage. In order to partially remove the water from the article, after the molding process has been completed the mold must be quickly aired.

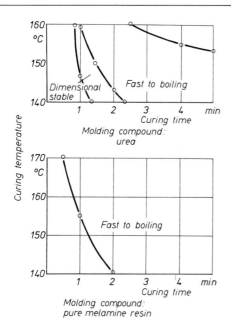

Figure 26
Curing time diagram. Molding compound pressed at room temperatures, wall thickness 2.5 mm [13].

Besides stearic acid (or its salts and amides), reactive plasticizers are used as flow agents for amino resins. Examples are:

Glycerol [14]

$$\underset{CH_2}{\overset{OH}{|}} - \underset{CH}{\overset{OH}{|}} - \underset{CH_2}{\overset{OH}{|}}$$

p-Toluene sulfamide [15]

$CH_3-\!\!\big\langle\!\!\_\!\!\big\rangle\!\!-SO_2-NH_2$

Monocresyl glycerol ether

$\underset{CH_2}{\overset{OH}{|}}-\underset{CH}{\overset{OH}{|}}-CH_2-O-\!\!\big\langle\!\!\_\!\!\big\rangle$
$\qquad\qquad\qquad\qquad CH_3$

Diamine

NH_2-R-NH_2

Alkylol amine

NH_2-R-OH

The alcohol and the amino group are able to react with the methylol groups. Since these reactions proceed only very slowly at low temperatures, the reaction between resin and plasticizer sets in essentially after the molding process is completed. After completing its primary function as plasticizer, the plasticizer reacts and thereby modifies the end products.

During the curing of amino resins, as in to the curing of phenolic resins, heat release and reaction shrinkage are slight since the curing is caused by few molar conversions.

During the curing of amino resins, as in the curing of phenol resols, water is eliminated and will be retained by the resin matrix and the filler. Hydrophilic fillers bind the water by

Table 32   Composition of total volatiles of melamine resin molding compounds during tempering (130 °C, 24 hr) according to [16]

Substance	% by mole	% by weight
H_2O	92.01	86.62
H_2	0.10	0.01
$N_2 + CO$	1.17	1.71
CO_2	1.41	3.25
CH_3OH	0.74	1.24
CH_2O	4.57	7.17

physicochemical bonds; water in the resin is presumably embedded in the micropores. Tempering of the articles results in loss of micropores and the volatile products escape. According to [16] these products are as listed in Table 32.

Tempering produces a simultaneous reduction in volume and weight; the percent reduction by weight is higher (Figure 27).

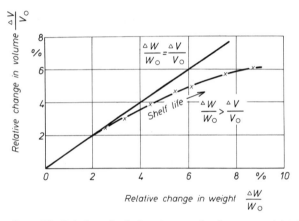

Figure 27   Relation of relative change of volume to weight during tempering of a melamine resin at 110 °C [16].

7.5.7   Urea Formaldehyde Foams

In principle, the same reactions occur during the synthesis of urea formaldehyde foams as during the curing of molding compounds. The resin composition is adapted to the specific processing method.

During the synthesis of foam, initially, an aqueous solution of a foaming agent and oxalic acid as condensation catalyst, solution (a), are mixed with an aqueous solution of a urea formaldehyde condensate, solution (b), in a mixing chamber. Gassing with air will produce a microcelluar foam. In the beginning, the foam is stabilized by surface tension only. In addition to the starting compounds, the mixture contains resorcinol as a highly reactive component and free formaldehyde [17]. At room temperature, the activator facilitates rapid condensation, which is accelerated particularly by resorcinol. Further condensation of the resin results in a rigid cell structure which stabilizes the foam structure. The foam is separated from water by progressive condensation. Finally, force of gravity causes the adsorbed water to run out. Since the polymer is a genuine open-cell foam, the water is allowed to run out freely.

Since, in addition to urea formaldehyde condensates, resorcinol is used as raw material, foams made in this manner are not pure amino resins. Alkylnaphthalene sulfonic acids are used as foaming agents. Their foaming power is very high; they are incorporated into the macromolecule during the reaction [18]. If tert.-butylnaphthalene sulfonic acid is used, then the tert.-butyl group increases (by inductive effect) the reactivity of naphthalene so much that a reaction with formaldehyde is possible. The reaction of tert.-butylnaphthalene sulfonic acid shows similarities to the synthesis of phenolic resins.

t-Butylnaphthalene-   Methylolcompound of
sulfonic acid   t-Butylnaphthalene sulfonic acid

Bibliography to Section 7.5

[1] *Ullmann, F.:* »Encyclopädie der technischen Chemie«, Bd. 3, 3. Auflage, S. 475–496, Verlag Urban und Schwarzenberg, München-Berlin, 1953.

[2] *Vale, C. P.:* »Aminoplastics«, Cleaver Hume Press, Ltd., London, 1950.

[3] *Blais, I. F.:* »Amino Resins«, Reinhold-Publishing Corp., New York, 1959.

[4] *Vieweg, R., E. Becker:* »Duroplaste«, Kunststoff-Handbuch, Bd. X, Carl Hanser Verlag, München, 1968.

[5] Lit. [4], S. 342.

[6] *Kulmin, K., U. Simonson:* Angew. Makromol. Chem. 68 (1978) 175.

[7] *Braun, D., V. Legradic:* Angew. Makromol. Chem. 34 (1973) 35.

[8] Lit. [4], S. 160.

[9] *de Jong, J. I., J. de Jong:* Rec. Trav. Chim. Pays-Bas 71 (1952) 643.

[10] *Zigeuner, G.:* Fette, Seifen, Anstrichmittel 57 (1955) 100.

[11] *Staudinger, H., K. Wagner:* Makromol. Chem. 15 (1955) 75.

[12] Lit. [4], S. 160.

[13] Lit. [4], S. 375.

[14] *Doehlemann, E., H. G. Knoblauch:* Kunststoff Rundschau 2 (1955) 37.

[15] *Laurie, W. A.:* SPE Journal 17 (1961) 31.

[16] *Behmer, A.:* Dissertation at the Institut für Kunststoffverarbeitung, RWTH Aachen, 1975.

[17] *Baumann, H.:* Kunststoffe 69 (1979) 440.

[18] *Garvey, M. J., Th. F. Tadros:* Kolloid Z. Z. Polymer 250 (1972) 967.

8 Production of Parts Using a Polyaddition Reaction

Polyaddition is the formation of polymers by addition reaction of two or more functional components without the liberation of any cleavage products. In principle, only two kinds of polyadditions are used in the polymer processing plant:

 – Addition onto epoxy groups (epoxy resins)
 – Addition onto isocyanate groups (polyurethanes)

Epoxy resins used for the manufacture of articles are made by addition of compounds with reactive hydrogen atoms (carboxyl, hydroxyl, amino groups) to epoxy groups. Polyurethanes are made by addition of hydroxyl compounds onto isocyanates; sometimes amino compounds are used together with hydroxyl compounds; the result is the formation of polyureas in addition to polyurethanes. Besides adding onto compounds which have reactive hydrogen atoms, isocyanate groups are able to react with themselves. This reaction is used for the synthesis of polyisocyanurates. The isocyanate group is also able to react with itself and form a carbodiimide group by eliminating carbon dioxide as by-product. Likewise, the synthesis of polycarbodiimides is used as a technical process; it is not a polyaddition but a polycondensation reaction. Since the formation of carbodiimide is a special process belonging to the chemistry of isocyanate compounds, it will not be discussed under polycondensation but in connection with the polyaddition of isocyanates.

Polyaddition, like polycondensation, consists of independent reaction steps. The oligomer or polymer compounds formed by these separate reactions have functional end groups which exhibit the same reactivity as the starting compounds. Initially, the increase in the viscosity of the resins is slow and the resins exhibit fluidity even at a high degree of conversion. While it is possible to cure low reactive epoxy resins in several stages, which means that cooling will stop these reactions even after a certain degree of conversion has been reached, it is practically impossible to interrupt the synthesis of polyurethane.

8.1 Polyurethanes

Polyurethane (PUR) is the classification of polymers which form a urethane group during the last step of the synthesis:

$$R_1 - NH - \underset{\underset{O}{\parallel}}{C} - O - R_2 \qquad \text{Urethane group}$$

The synthesis of polyurethane starts with polyisocyanates and polyols; "poly" means more than one. Polyisocyanates and polyols are liquids at processing temperatures.

Although PUR is only recently developed compared to the other polymers [1], its adaptability to many diverse uses has made it one of the most abundantly produced volume synthetic polymers today.

In 1978, West Germany used 224,000 tons of PUR; 5% of that was thermoplastic PUR. The remaining 95% was synthesized by the polymer processor during the molding process.

The true PUR producer is not the chemical industry but the various polymer processing plants.

85% of PUR is used as foams [3]. It is not the finished polymer that is foamed but a reactive mixture of liquid prepolymers. The heat of reaction can be utilized to evaporate the blowing agents. The synthesis, processing, and properties of the polyurethanes and their prepolymers are described in detail in [4–9].

8.1.1 Synthesis of PUR in the Polymer Processing Plant

Polyol and polyisocyanate, the starting compounds for the synthesis of PUR, are relatively stable components with limited shelf life. Both components are sensitive to moisture. Polyols absorb humidity, while polyisocyanates react with water spontaneously. The basic components are mixed prior to the molding process. The formation of PUR starts during the mixing process. The use of a catalyst can reduce the curing time of the resin to less than a minute. The high rate of reaction is of particular importance to the economical feasibility of PUR synthesis; but places high demands on the processing plant. The raw materials must be stored in separate containers. The additives must always be added to the polyol compound; the isocyanates are used as is.

Effective and rapid mixing of the prepolymers is necessary for the processing of PUR articles. Since, for fast reacting systems, gelation of the resin sets in very early, the mixing of the components has to take place in a fraction of a second. Highly reactive mixtures are reacted in small mixing chambers using the injection ribbon mixer process. Intermittent operation leads to curing of the resin in the mixing chamber. Therefore, the mixing chamber either has to be part of the outlet or it has to be cleaned with compressed air or by solvents.

Adherence to constant ratios of the mixture presents an additional problem for the PUR processor; an accuracy of $\pm 1\%$ is necessary.

8.1.2 Raw Materials for PUR Production

In addition to the reactants (polyol and polyisocyanate), additives have to be used, and a catalyst is necessary to accelerate the polyaddition. A multitude of various catalysts with diverse effectiveness are available. Catalysts are often called "activators". Resins used for the synthesis of PUR may contain, in addition to catalysts, anti-aging additives, antioxidants, plasticizers, dyes and fillers. Addition of a blowing agent (usually a low-boiling halogenated hydrocarbon) and foam stabilizers is necessary for the manufacture of foams. The stabilizer stabilizes the bubble structure formed by the low viscous resin until cell rigidity sets the structure of the foam.

(a) Polyisocyanates

Only a few polyisocyanates are commercially available: these are essentially the diisocyanates of toluene and of diphenyl methane.

Two isomers of toluylene diisocyanate (TDI) exist. The different species of TDI are based on the different ratios of the two isomers. For instance, ratios of 2,4 TDI to 2,6 TDI of 100:0, 80:20, and 60:40 are customary commercial for products. Since the 2,4 isomer is the more reactive one, the isomer ratio is of technical importance.

CH₃

NCO   2,4-Toluylene diisocyanate
       (2,4-TDI)

NCO

CH₃

OCN   NCO   2,6-Toluylene diisocyanate
              (2,6-TDI)

Diphenyl methane diisocyanate (methylene diphenyl 4,4'-diisocyanate), abbreviated MDI, is another important polyisocyanate.

OCN—⟨ ⟩—CH₂—⟨ ⟩—NCO

Pure MDI is crystalline (m.p. 39.4 °C) at room temperature, and its shelf life is poor. Cyclotrimerization converts it into an inactive isocyanurate. Most of the commercially available products can be classified into two groups:

1. The MDI, obtained by synthesis, is reacted with small quantities of hydroxyl compounds; so that, a small number of isocyanate groups are converted beforehand. In principle, prepolymers are being formed. This modification has technical advantages. It will further lower the MDI vapor pressure which is already quite low at room temperature, allowing the crystallization of MDI at even lower temperatures. At the same time, MDI becomes more stable since the possibility of NCO-consuming reactions (trimerization) is reduced. Modifying agents, for example, are low molecular alcohols. The use of triols will yield higher functional isocyanates.

2. Another reason for the multitude of commercially available types of MDI is stipulated by the synthesis. MDI is synthesized by reaction of aniline with formaldehyde and subsequent reaction of the resulting diamines with phosgene.

Since the reaction of aniline with formaldehyde leads to different reaction mixtures depending on the condition of the process, the resulting types of crude MDI will be different and will lend their specific properties to the articles made by PUR synthesis. Besides various isomers, higher functional isocyanates are formed.

Isomers formed by the reaction between aniline and formaldehyde can be isolated individually. The reaction of pure isomers will result in various pure diisocyanate compounds. A separation of the isomers is not necessary. Crude MDI consists of 4,4' isomers and various other di- and triisocyanates.

Besides TDI and MDI, several other diisocyanates are used to a small extent; for example:

ONC – CH₂– CH₂– CH₂– CH₂– CH₂– CH₂ – NCO   Naphthalene- 1,5-diisocyanate

NCO

OCN   Hexamethylene diisocyanate

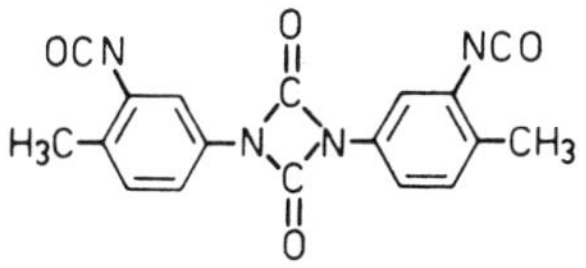

N,N'-(4,4"-Dimethyl-3,3"-diisocyanate-diphenyl)-
uretdione (Uretdione made from TDI)

Hexamethylene diisocyanate is used in the synthesis of PUR elastomers and yellow-ing-resistant PUR coatings. Uretdione, made from TDI, is a diisocyanate which reacts with diol and produces a linear (thermoplastic) PUR containing uretdione groups. Cross-linkage of the polymer is made possible by the uretdione groups (see Section 10.5). The polyisocyanates are characterized by their percentage of isocyanate. The amount of isocyanate is stated in percent by weight. An isocyanate content of 32 means that 100 g of polyisocyanate contains 32 g of NCO groups (MW = 42), or correspondingly:

$$\frac{32 \text{ g} \cdot \text{mol}}{42 \text{ g}} = 0,76 \text{ mol NCO groups}$$

A quantitative analysis of the isocyanate content in the compound is obtained by reacting the isocyanates with di-n-butylamine or diisobutylamine in a solution of chlorobenzene. This reaction causes the isocyanate to become trisubstituted urea. After the reaction is completed the solution is diluted with methanol and the excess amine is titrated with aqueous hydrochloric acid. The calculated number of moles of amine consumed is equal to the number of moles of isocyanate groups [10].

(b) Polyols

The multitude of different polyurethanes are based on the large number of different polyols that are available for PUR synthesis. Polyols that are used commercially for PUR synthesis form a separate class:

Polyols which are suitable for PUR synthesis include both polyols based on natural products and polyols prepared synthetically.

As a proven natural polyol, castor oil has been proven successful, it contains three available hydroxyl groups which produce cross-linked polymers.

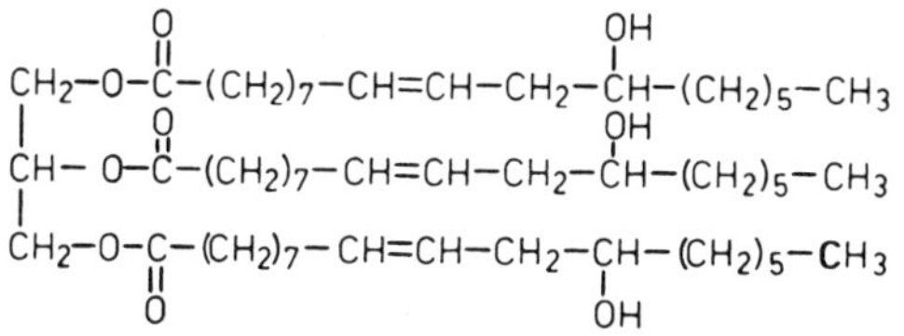

The literature describes the use of tall oil, which is a by-product in the manufacture of cellulose, for the synthesis of PUR [4]. In principle, the following polyols can also be used: sugar, starch, cellulose, and lignine. These compounds are solids, and therefore they cannot produce homogeneous polyurethanes, and they are suitable only as reactive fillers. To be used for PUR synthesis, they must be converted chemically into liquid compounds.

The synthetically prepared polyols are much more important. They are divided into two groups:

- Polyester polyols
- Polyether polyols

Table 33   Starting materials for the synthesis of polyester polyols

Formula	Name
HOOC – (CH$_2$)$_4$– COOH	Adipic acid
 HOOC——⬡——COOH	Phthalic acid
HOOC – (CH$_2$)$_8$ – COOH	Sebacic acid
HO – CH$_2$– CH$_2$ – OH	Ethylene glycol
HO – CH$_2$– CH$_2$– O – CH$_2$– CH$_2$ – OH	Diethylene glycol
HO – CH$_2$– CH$_2$– O – CH$_2$– CH$_2$– O – CH$_2$–CH$_2$– OH	Triethylene glycol
$\overset{\displaystyle CH_3}{\underset{\displaystyle \vert}{}}$ HO – CH$_2$–CH – OH	1,2-Propylene glycol
$\overset{\displaystyle OH}{\underset{\displaystyle \vert}{}}$ $\overset{\displaystyle OH}{\underset{\displaystyle \vert}{}}$ CH$_3$– CH – CH$_2$– O – CH$_2$– CH – CH$_3$	Dipropylene glycol
$\overset{\displaystyle OH}{\underset{\displaystyle \vert}{}}$ HO – CH$_2$– CH$_2$– CH – CH$_3$	1,3-Butylene glycol
HO – CH$_2$– CH$_2$– CH$_2$– CH$_2$ – OH	1,4-Butanediol
HO – (CH$_2$)$_6$ – OH	1,6-Hexanediol
$\overset{\displaystyle CH_3}{\underset{\displaystyle \vert}{}}$ HO – CH$_2$–C – CH$_2$– OH $\underset{\displaystyle CH_3}{\vert}$	Neopentyl glycol
OH OH OH $\vert$ $\vert$ $\vert$ CH$_2$– CH – CH$_2$	Glycerin
$\overset{\displaystyle C_2H_5}{\underset{\displaystyle \vert}{}}$ HO – CH$_2$–C – CH$_2$– OH $\underset{\displaystyle CH_2 – OH}{\vert}$	Trimethylolpropane
$\overset{\displaystyle OH}{\underset{\displaystyle \vert}{}}$ HO – CH$_2$– CH – CH$_2$– CH$_2$– CH$_2$– CH$_2$ – OH	Hexane triol-1,2,6
HO – CH$_2$ ⟍ ⟋ CH$_2$ – OH C HO – CH$_2$ ⟋ ⟍ CH$_2$ – OH	Pentaerythritol

Polyester Polyols

Polyester polyols are made by esterification of diols with dicarboxylic acids using an excess of alcohol. Bifunctional compounds lead to linear polyester polyols. Using, or adding, compounds containing more than two functional groups results in branched polyols. Cross-linked polyols are unsuitable for PUR processing; they do not deliver homogeneous polyol isocyanate mixtures. Molecular weights of polyester polyols are between 400 and 4000.

The complete removal of the water formed during synthesis of polyester polyol is not economically possible; the user of polyol has to take that (the reaction with the water, see page 138) into consideration. Therefore, it is necessary to quote the percentage of water for each polyol. During synthesis of high molecular weight polyester polyols, a small percentage of carboxyl groups remains in the molecule. This cannot be avoided, and these carboxyl groups too will react with isocyanate.

Polyether Polyols

Polyether polyols are polymerization products of cyclic oxides (see Table 34).

Table 34   Starting materials for the synthesis of polyether polyols

Formula	Name
$\overset{O}{\overset{\diagup\diagdown}{CH_2-CH_2}}$	Ethylene oxide
$\overset{O}{\overset{\diagup\diagdown}{CH_3-CH-CH_2}}$	Propylene oxide
$\overset{CH_2-CH_2}{\underset{CH_2-CH_2}{\vert}}\!\!\searrow\!\!O$	Tetrahydrofuran
$H_2N-CH_2-CH_2-NH_2$	**Starter:** Ethylenediamine
$H_2N-CH_2-CH_2-NH-CH_2-CH_2-NH_2$	Diethylenetriamine
$CH_3-\!\!\overset{NH_2}{\underset{}{\bigcirc}}\!\!-NH_2$	Aromatic amines
$HO-CH_2-CH_2-OH$	Ethylene glycol
$HO-CH_2-\overset{OH}{\underset{\vert}{CH}}-CH_2-OH$	Glycerin

Usually an alcohol is used to initiate polymerization; diol results in a linear polyol, and triol in a branched polyol. Polyether polyol and polyester polyols are only partially miscible with each other; even alcohols like butane diol are not very compatible with polyols.

Polyester polyols produce polyurethanes which differ in their thermal properties and tolerance to corrosive substances from the polyurethanes produced by polyether polyols. Polyether polyols are thermally more unstable and more easily oxidized than polyester polyols, but they are more stable to saponification.

Polyols are characterized by their hydroxyl number (OH number), which is defined as the amount of potassium hydroxide (100%) (in mg) which is needed to saponify 1 g of completely esterified polyol. The first step in the quantitative analysis is esterification of the polyol with acetic anhydride using pyridene as solvent. Phthalic anhydride is used instead of acetic anhydride in the analysis of polyether polyols. Free acid formed by esterification is titrated with 0.1 N aqueous sodium hydroxide; the so determined number of moles of sodium hydroxide corresponds to the number of moles of OH groups in the polyol [11]. The number of carboxyl groups in a polyol is determined by direct titration (of the polyol) in aneutral solution of ethanol and benzene (1:1) with 0.1 N aqueous sodium hydroxide; specification is in mg equivalent KOH per gram of polyol [12].

A blend of various polyols is generally used for the processing of PUR instead of using individual polyols. PUR polymers processed from a mixture of polyols have better properties. Commercially available polyols are mostly mixtures of polyols. In principle, the processor is able to prepare his own polyol mixture, which then will result in a polyurethane polymer with optimal properties for a specific use.

8.1.3 Chemistry of the Isocyanate Group

The isocyanate group is a very reactive functional group and can be used in a variety of addition reactions. The most important components for addition reaction with the isocyanate group are compounds which contain "active" hydrogen atoms which are easily exchangeable. The mechanism of these reaction is as follows:

$$R_1 - N = C = O \quad + \quad H - R_2 \quad \longrightarrow \quad R_1 - NH - \overset{\overset{\displaystyle O}{\|}}{C} - R_2$$

The compound H-R represents a variety of products: polyol, filler, end product, or impurity.

The ability of the $-N = C = O$ group to react with itself is a very important property used in polymer processing. The polyisocyanurate process makes use of this reaction because formation of isocyanate chains will increase the heat distortion temperature of polyurethane.

During PUR synthesis, the most important reaction of the isocyanate group is the addition of alcohols. The resulting product is the urethane group. Other reactions are also important during PUR synthesis, especially those with

- Water
- Amines or amides
- Carbonic acids
- Urethanes
- Urea derivatives

These reactions are listed in Table 35. Side reactions b) to f) occur to only a small extent. Synthesis of nonrigid foam uses large amounts of water as reactant; which makes several side reactions quite important.

Water may be residual water from the synthesis of polyol or it may have been introduced subsequently as a contamination. Gaseous carbon dioxide is formed as a result of

Table 35 Important reactions of the isocyanate group

a)	$R^1-N=C=O$ + $HO-R^2$ → $R^1-NH-\overset{\overset{O}{\|\|}}{C}-O-R^2$ Alcohol → Urethane
b)	$R^1-N=C=O$ + $H-OH$ → $R^1-NH-\overset{\overset{O}{\|\|}}{C}-OH$ Water → Carbamic acid Secondary reaction: $R^1-NH-\overset{\overset{O}{\|\|}}{C}-OH$ → CO_2 + R^1-NH_2 Amine
c)	$R^1-N=C=O$ + $H-NH-R^2$ → $R^1-NH-\overset{\overset{O}{\|\|}}{C}-NH-R^2$ Amine → Urea derivative
d)	$R^1-N=C=O$ + $HOOC-R^2$ → $R^1-NH-\overset{\overset{O}{\|\|}}{C}-O-\overset{\overset{O}{\|\|}}{C}-R^2$ Carboxylic acid → Mixed anhydride Secondary reaction: $R^1-NH-\overset{\overset{O}{\|\|}}{C}-O-\overset{\overset{O}{\|\|}}{C}-R^2$ → $R^1-NH-\overset{\overset{O}{\|\|}}{C}-R^2$ + CO_2 Amide
e)	$R^1-N=C=O$ + $R^2-O-\overset{\overset{O}{\|\|}}{C}-NH-R^1$ → $R^2-O-\overset{\overset{O}{\|\|}}{C}-\underset{\underset{R^1}{\|}}{N}-\overset{\overset{O}{\|\|}}{C}-NH-R^1$ Urethane → Allophanate
f)	$R^1-N=C=O$ + $HN-\overset{\overset{O}{\|\|}}{\underset{\underset{R^1}{\|}}{C}}-NH-R^2$ → $R^1-NH-\overset{\overset{O}{\|\|}}{C}-\underset{\underset{R^1}{\|}}{N}-\overset{\overset{O}{\|\|}}{C}-NH-R^2$ Urea → Biuret

reaction with water. During the processing of solid articles, formation of CO_2 is undesirable, because it can produce voids in the article; therefore, it is necessary to eliminate water from the prepolymer. The hydrophilicity of polyols makes the complete elimination of water extremely difficult. Water-absorbent additives like zeolites are used occasionally to correct this problem. On the other hand, the formation of voids can be prevented by using pressure. Carbon dioxide is soluble in PUR under pressure, and increasing pressure by less than 10 bar during the curing process of the polymer is usually sufficient. The controlled formation of carbon dioxide by intentional addition of water is of commercial importance. The generated carbon dioxide acts as an expanding agent and will expand the curing resin.

Another product formed by the "water reaction" is an amine which reacts faster than polyols with isocyanate. Addition of amine onto the isocyanate group results in a urea derivative. Formation of urea is a necessary consequence of the "water reaction." In total

1 mole of water uses 2 moles of isocyanate groups. One gram of water uses 19.33 g of toluylene diisocyanate or 27.78 g of (pure) MDI. Formation of the polymer is not adversely affected if the calculation of the ratio of ingredients takes the percentage of water into consideration. Instead of a pure polyurethane, a polyurethane-urea polymer will be produced.

Carboxylic acids add to isocyanate groups and form mixed anhydrides which are derivatives of a carboxylic and a carbamic acid. These anhydrides are unstable at higher temperatures and decompose into amide and carbon dioxide. Like water, carboxylic acids will cause the formation of carbon dioxide. Usually, in the polyester polyols a few carboxylic groups remain which have not been esterfied during polyol synthesis.

Urethane and urea are very reactive compounds themselves, since they too have an exchangeable hydrogen atom. The same is true of the amide group. Polyurethanes and polyureas can also add to isocyanate groups. The urethane group will form an allophanate group, and the urea group is converted to a biuret group.

Formation of allophanate is limited, since urethane groups react more slowly than alcohols with isocyanate groups. It is characteristic for the formation of allophanate to generate branching or cross-linkage positions. This is not an important factor in the synthesis of highly cross-linked polyurethanes, but polyurethanes that are only slightly cross-linked will show an increase in cross-link density.

Urea groups react faster than urethane groups with isocyanate. Only foams that have been expanded with water contain large amounts of urea groups and only here is the formation of biuret a significant factor.

Another interesting reaction is the addition of hydroperoxides to isocyanates:

$$R_1 - N = C = O \quad + \quad H - O - O - R_2 \longrightarrow R_1 - NH - \overset{\overset{\displaystyle O}{\|}}{C} - O - O - R_2$$

$$\text{Hydroperoxide} \qquad \text{Peroxyurethane}$$

The use of thermally stable hydroperoxides will result in spontaneous conversion. The hydroperoxy group reacts with the isocyanate molecule and produces a new peroxide, peroxyurethane. The newly formed peroxide decomposes at a lower temperature than the hydroperoxide, and a radical generator which decomposes at low temperatures is formed in situ.

Important reactions of the isocyanate molecule are addition reactions (cyclodimerization and trimerization) and condensations which produce carbodiimides. These reactions produce polymers like polyisocyanurate and polycarbodiimide and are summarized in Section 8.2.

8.1.4   Catalysis of Urethane Formation

A mixture of pure polyisocyanate and pure polyol reacts very slowly at first. After the conversion of only a small portion of the reactants, the rate of reaction increases, since the reaction product accelerates the formation of urethane (auto catalysis). Uncatalyzed reactions are too slow for the manufacture of parts made from PUR; hence the general use of catalysts.

Polyurethanes are formed by the addition of a negatively charged oxygen atom of the polyol onto the carbon atom of the isocyanate group which contains a positive partial charge.

$$R_1 - N = C = O \quad \text{Isocyanate}$$
$$\overset{\delta+}{}$$
$$H \overset{\delta-}{-} O - R_2 \quad \text{Alcohol (Polyol)}$$

A rapid formation of urethane is the result of highly positively or negatively charged particles. The space requirements of the radicals R_1 and R_2 (steric hindrance) and the mobility of the hydrogen atom in the hydroxyl group are also significant.

Two groups of catalysts (accelerators) will accelerate the formation of polyurethane:

- Basic catalysts
- Metal compounds

One result of the effect of the basic catalysts may be the promotion of hydrogen transfer in an intermediate compound:

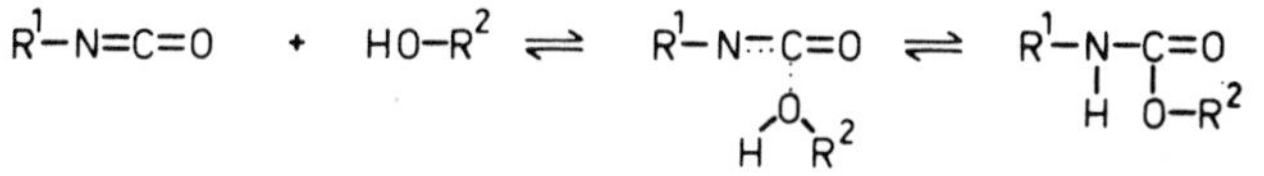

$$R^1-N=C=O \quad + \quad HO-R^2 \rightleftharpoons R^1-N \overset{..}{=} C=O \rightleftharpoons \overset{R^1-N-C=O}{\underset{H \quad O-R^2}{|}}$$
$$\overset{O}{\underset{H \quad R^2}{}}$$

Intermediate compound

Some basic catalysts are listed in Table 36. Tertiary amino bases act only as accelerators. In addition to that, accelerators exist which are incorporated into the polymer through an amino group or another functional group. If the accelerator molecule has more than one

Table 36 Basic catalysts for the formation of PUR

Formula	Name
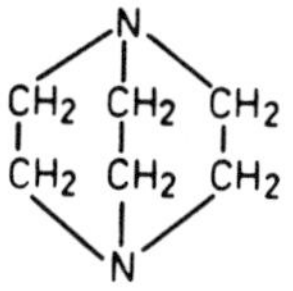 ![triethylene diamine structure] CH₂ CH₂ CH₂ / CH₂ CH₂ CH₂ with N at top and bottom	Triethylene diamine (DABCO)
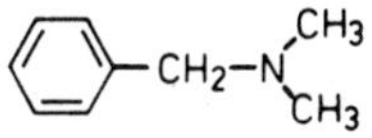 benzyl ring—CH₂—N(CH₃)(CH₃)	Dimethyl benzylamine
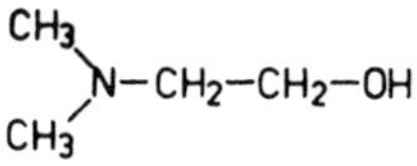 (CH₃)(CH₃)N—CH₂—CH₂—OH	Dimethyl aminoethanol
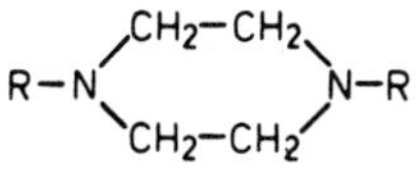 R—N(CH₂—CH₂)(CH₂—CH₂)N—R	Piperazine
H₂N—CH₂—CH₂—NH₂	Ethylenediamine
H₂N—CH₂—CH₂—NH—CH₂—CH₂—NH—CH₂—CH₂—NH₂	Triethylenetetramine
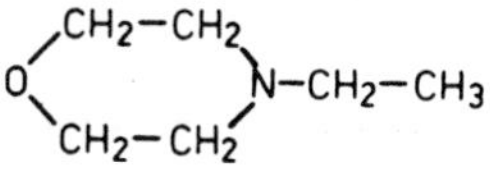 O(CH₂—CH₂)(CH₂—CH₂)N—CH₂—CH₃	Ethylmorpholine

group which is able to react with isocyanate, then the accelerator will also act as a cross-linking agent. Ethylenediamine is a very effective accelerator which is incorporated into the polymer; the resulting urea groups are basic and therefore able to act as an accelerator.

Organic compounds containing tin (for example, dibutyl tin dilaurate) are the most important metal catalysts used in the formation of PUR. They differ from the basic catalysts by having a substantially higher activity. One possible mechanism is the formation of a complex consisting of a catalyst and an isocyanate group, where the isocyanate group shows a substantially higher activity [13]. A few catalytic metal compounds are listed in Table 37.

Table 37   Metal Catalysts for PUR Synthesis.

Formula	Name
$C_4H_9 - Sn \; Cl_3$	Butyl-tin-IV chloride
$Sn \; Cl_2$	Tin-II chloride
$Sn \; (C_7H_{15} - COO)_2$	Tin octoate
$Sn \left[CH_3 - (CH_2)_7 - CH = CH - (CH_2)_7 - COO \right]_2$	Tin oleate
	Dibutyl-Tin-diocytyl maleate
$(C_4H_9)_4 \; Ti$	Tetrabutyl titanate
$Fe \; Cl_3$	Ferric chloride
$Sb \; Cl_3$	Antimony-III chloride
	Cobalt 2-ethylhexoate

Metal catalysts containing long-chain carboxylic acids have the disadvantage of being hydrolyzed by traces of water present in the polyol [14].

Dibutyl-tin-dilaurate   Dibutyl-tin(IV)-   Lauric
                             oxide   acid

Mixtures of polyol and catalyst have limited shelf life because of their sensitivity to hydrolysis.

Since the mechanisms are different for base and metal accelerators, it is possible to use both kinds of catalysts simultaneously, resulting in a synergistic effect.

Sometimes it is necessary to slow down the formation of urethane during PUR synthesis. Inhibitors for this reaction are acidic compounds such as phosphoric acid, organic acid halides, or weak acids such as benzoic acid [5]. Their effectiveness is due to the neutralization of alkaline groups that exhibit catalytic activity. Inhibiting the formation of urethane will result in greater leeway in the processing of resins.

8.1.5 Reactivity of Reactants

The reactivity of an isocyanate depends on the structure of the remainder of the molecule. The reactivity of the two isocyanate groups for symmetrical diisocyanates is the same, but after one isocyanate group has reacted, the reactivity of the other group is slightly altered. This means that even symmetrical diisocyanates show a gradation in the reactivity of both isocyanate groups (see MDI, see page 132).

In TDI, an isocyanate group in the para position is more reactive than in the ortho position. Initially, 2,4 TDI reacts with alcohol in the para position:

The reactivity of the isocyanate group in the ortho position depends on the methyl group and the substituent in the para position (to the methyl group). Altering the substituent in the para position will automatically alter the reactivity of the isocyanate group in the ortho position.

A summary of values showing the different reactivities of the two isocyanate groups in diisocyanates is given in Table 38. The reactivity of the isocyanate group has the highest value at the beginning of the reaction. The increasing bulk temperature causes an increase in the rate of reaction which, in actual use, more than compensates for the loss of reactivity during the reaction. Elevation of temperature, however, does not alter the relative reactivities of the two isocyanate groups in a given molecule.

Table 38 Relative rate of reaction and activation energies of the reaction of isocyanates with excess ethanol in toluene at 30 °C [15]

Isocyanate	Relative rate of reaction		Activation energies kJ	
	1. NCO-group	2. NCO-group	1. NCO-group	2. NCO-group
Isocyanatobenzene	14	–	40	–
2,4-Diisocyanate-toluene	42	1.7	46	–
4,4′Diisocyanate-diphenylmethane	32	10	30	39
1,5-Diisocyanate-naphthalene	38	11	21	39
ω-Isocyanate-toluene	1.4	–	30	–
ω,ω′-Diisocyanate-m-xylene	2.8	1.1	53	50
ω,ω′-Diisocyanate-p-xylene	2.5	1.3	58	47
5-tert.-Butyl-ω,ω′-diisocyanate-m-xylene	2.7	1.0	49	50

At the beginning of the reaction, the gradation in reactivity causes only one side of the isocyanate component to react; therefore, slowing down the increase in viscosity occurs only gradually. In MDI, gradation in the reactivity of isocyanate groups is hardly noticeable; on the other hand, in 2,4-TDI it is quite high (difference in reactivity is 1:25). Plotting the logarithm of the concentration of unreacted isocyanate against reaction time should result in a straight line, since this is a second-order reaction (for reaction in solvents). However, the dependence is not linear on a logarithmic scale. Deviation from a straight line is greater for TDI than for MDI (see Figure 28).

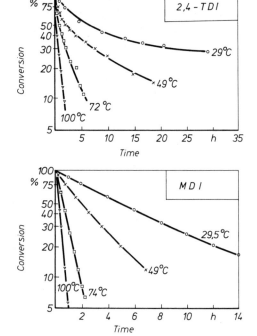

Figure 28
Reactions of 2,4-toluene
diisocyanate (top) and
4,4'-diphenylmethane diisocyanate
(bottom) with diethylene glycol
(in chlorobenzene) [5, 16].

The reactivities of primary alcohols are substantially higher than those of secondary alcohols. Tertiary alcohols react very slowly and tend to decompose, forming olefins.

In polyols, or mixtures of polyols having groups with different reactivities, the primary hydroxyl group is the first group to react [17]. This behavior can be used to control the overall reaction. For instance, if a mixture of a slow-reacting, long-chain polyol and butane diol is used, butane diol will react first, forming rigid segments. The long-chain polyol reacts slowly and results in segments with a low number of urethane groups (elastomeric segment).

Table 39   Relative reactivities of alcohol toward isocyanates [18]

Alcohol	Relative reactivity
Methanol	1.2
1-Butanol	1.4
Isopropyl-alcohol	0.6
Ethylene glycol monomethylether	0.14

8.1.6 Ratio of Polyol to Polyisocyanate

In theory, a resin mixture of equimolar amounts of hydroxyl and isocyanate groups should result in a polymer with optimal properties. In practice, resins with equimolar amounts of reactants are seldom used. Optimal qualities are obtained by using an excess of molar amounts of isocyanate groups.

For the manufacture of elastomeric polyurethanes, only a slight excess of isocyanate is used. A small, defined percentage of cross-linkage is characteristic of the elastomeric polyurethanes. An excess of isocyanate can lead to a side reaction and the formation of allophanate (see eqaution e, Section 8.1.3), resulting in further cross-linkage. It is important, therefore, to adhere to an exact ratio of components; a very slight excess of isocyanate (1%) is necessary to compensate for contaminations which may react with isocyanate. The use of equimolar amounts of isocyanate and hydroxyl groups is very effective and results in PUR with a very high molecular weight. An excess of 3–10 mole % of isocyanate is needed for the manufacture of rigid PUR (PUR rigid foam). Any isocyanate groups not used up by side reactions during the manufacturing process are converted, by water that penetrates the article, into ureas or amines.

Resin mixtures which use a catalyst to cyclotrimerize isocyanate groups to isocyanurate are able to handle any amount of excess isocyanate (see Section 8.2.3). Excess isocyanate reacts simultaneously to the synthesis.

The ratio of the two reactants is defined as the characteristic number; it expresses the molar amount of isocyanate groups as a percentage of the molar amount of hydroxyl groups. A characteristic number of 110 signifies an excess of 10 mole % of isocyanate groups. The polymer processor is able to calculate the ratio of polyol to polyisocyanate (if this is not given by the raw material manufacturer) by using the OH number and the remaining amount of isocyanate in polyisocyanate. The following is a rule of thumb [8]:

- A polyol contains 0.0303% by weight of OH groups per unit of OH number
- 1 OH unit requires 0.075 g of NCO groups for an equimolar mixture

It follows that the formula for the amount of polyisocyanate (which contains X gram % of NCO) is:

$$\text{Required amount of isocyanate} = \frac{7.5 \cdot \text{OH-number}}{\text{x\% by weight NCO}} \left(\frac{\text{g of Polyisocyanate}}{\text{g of Polyol}} \right)$$
$$(\text{CN} = 100)$$

If other characteristic numbers are being used, the required amount of isocyanate (in grams) has to be multiplied by:

$$\frac{\text{CN}}{100}$$

8.1.7 Curing Reactions

In an adiabatic system the conversion of the reactive reactant is easy to monitor by measuring the temperature. The heat of reaction is proportional to the amount of converted groups. In practice, adiabatic systems are not available. Since the structure of the foam reduces the thermal conductivity, approximately adiabatic conditions prevail for reactions that take place in the core of the article during the manufacture of the foam. Thermal conductivity is higher inside the slid layers at the edges of the integral forms, and a larger percentage of the heat of reaction is lost to the surroundings [19].

Either the formation of PUR or the decrease in the concentration of isocyanate can be determined in test molds by infrared spectroscopy in combination with attenuated total reflection (ATR), a new technique [20]. Figure 29 illustrates temperature change and decrease in the concentration of isocyanate in different layers inside a test mold. Change in temperature is practically zero in the layer close to the wall of the mold because most of the heat is absorbed by the mold itself. On the other hand, the polymer in the center of the mold reaches higher temperatures depending on the exothermic reaction of the resin and the temperature of the walls of the mold. On the surface of the mold the chemical reaction occurs quite slowly because of the lower temperature; in the innermost part of the mold, which is quite hot, conversion of isocyanate groups takes place quite rapidly.

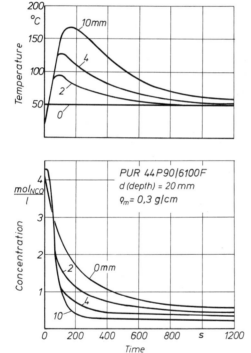

Figure 29
Concentration of NCO groups and bulk temperature in the test mold; measurements taken at various depths [20]; (44P90 and 6100F are commercial products made by Bayer AG).

At the beginning of the reaction, the viscosity of the resin increases only slowly (see Figure 30). Increase in viscosity depends not only on the type of base material but also on the temperature of the polymer mass. After a certain degree of conversion is reached, gelation of the resin sets in and it becomes meaningless to describe the resin by its viscosity. The shear modulus is a suitable parameter used to describe the degree of curing after the resin has began to gel.

The heat of reaction is a function of the concentration of reactive groups in the resin. It is approximately 105 kJ per mole of NCO (for equimolar amounts of NCO and OH) on the average [20, 21].

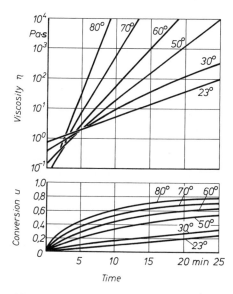

Figure 30
Isothermic viscosities measured and conversions calculated for manufacturing PUR foam [20].

Also, volume contraction during reaction is proportional to the concentration of converted groups. It is highest for polyols with a high OH number and isocyanate compounds with a high percentage of isocyanate groups.

Reaction of phenyl isocyanate with ethanol to form phenyl ethyl urethane is accompanied by 10.7% volume contraction. The contraction is 6.1% [22] during PUR synthesis from diisocyanate containing 28% (by weight) of NCO and a polyol with an OH index of 500.

In practice, reaction caused contraction during PUR synthesis is not important. The largest percentage of the contraction happens at a time when the polymer is still plastic, since cross-linkage (that is, fixation of molecules) takes place at a high degree of conversion only. The contraction is easily compensated for by the foaming pressure. After the geometric shape of the part is set, contraction caused by reaction is quite insignificant; this amount of contraction is added to the thermal shrinkage. The total deviation of the fabricated shape from the size of the mold depends on three factors:

– Contraction caused by reaction
– Thermal contraction
– Post-curing shrinkage

For foamed plastics, shrinkage caused by reaction is the least significant of these factors [23].

The slow rise in viscosity at the beginning of PUR formation is advantageous to the formation of foam. The temperature rises much faster than the viscosity. The heat of reaction can be used to evaporate the blowing agent, and the free-flowing foam easily fills even the most intricate molds. In exceptional cases, a quick, but limited, increase in viscosity is necessary at the beginning of the curing reaction in order to obtain the best foam structure. This is accomplished by the use of diamines with short chains, such as ethylenediamine (NH_2-CH_2-CH_2-NH_2). The amino group reacts faster with isocyanate than the hydroxyl group does; the short chain diamine (chain propagator) is easily used up and forms urea, thus increasing the viscosity of the resin. An amount of 0.1–1% by weight of ethylenediamine is enough to effectively increase the viscosity.

8.1.8 One-Component Reaction Casting

Since a mixture of a liquid polyisocyanate and a liquid polyol reacts spontaneously even at room temperature, the two reactants can be used only in a two-component reaction casting process. Therefore, it was proposed to use a mixture of a solid isocyanate and a liquid polyol for synthesis of PUR, in order to change the two-component process to a one-component process. This heterogeneous mixture does not cure until the isocyanate component is totally molten and therefore represents a thermosetting resin. For instance, the reaction mixture of 2 moles of toluylene diisocyanate and 1 mole of water does not melt below 178 °C; if polyol is added to the mixture, it melts between 150 and 160 °C. By adding tin compounds as catalysts, the starting temperature of the reaction can be lowered to 90–100 °C. The disadvantage of this process is the short shelf life (a few hours only) of these resins; a slow curing of the resin takes place despite the heterogeneity of the mixture.

Another process uses complex amines as chain propagators which will liberate reactive amines at higher temperatures only. These amines will react immediately with isocyanate to form urea [24].

One-component resins that are cured by addition of water are of practical use. These resins are prepolymers containing isocyanate groups. Water from the atmosphere converts the isocyanate groups to amino groups, eliminating carbon dioxide. These amino groups react immediately with the excess isocyanate to form urea. This formation of urea from amine and isocyanate causes chain propagation and chain linkage. The formation of urea is responsible for the curing process, and the resulting polymer is, strictly speaking, a polyurea, not a polyurethane.

Diffusion of water into the product is necessary for the reaction. It is impossible to run this process in a closed mold, since air (water) must be available for curing. This process depends on diffusion and is therefore suited for thin layers only. At the same time, formation of carbon dioxide takes place; this limits the process to the manufacture of foams.

Besides one-component resins which are cured by water, one-component resins with chemically shielded isocyanate groups are also used.

Urethanes synthesized from phenol will undergo thermal decomposition:

$$C_6H_5-O-\overset{\overset{O}{\|}}{C}-NH-(CH_2)_6-NH-\overset{\overset{O}{\|}}{C}-O-C_6H_5 \xrightarrow{170°C} \begin{array}{l} O=C=N-(CH_2)_6-N=C=O \\ + \\ 2\ C_6H_5-OH \end{array}$$

Resins with free hydroxyl groups and isocyanate groups shielded by phenol undergo exchange of the alcohol component with elimination of phenol when heated to 170 °C [4]. Since this is a condensation reaction, it cannot be utilized for the processing of molding compounds.

Uretdione groups represent another type of shielded isocyanate group. They are able to react with hydroxyl groups at higher temperatures. This reaction can be utilized to crosslink thermoplastic PUR (see Section 10.5).

Bibliography to Section 8.1

[1] *Bayer, O.:* Angew. Chem. 59 (1947) 257.

[2] *Hirtz, R., K. Uhlig:* Chem. Ind. 30 (1978) 617.

[3] *N. N.:* Kunststoffe 69 (1979) 496.

[4] *Vieweg, R., A. Höchtlen:* »Polyurethane«, Kunststoff-Handbuch, Bd. VII, Carl Hanser Verlag, München, 1966.

[5] *Saunders, J. H., K. C. Frisch:* »Polyurethanes, Chemistry and Technology«, Interscience Publishers, John Wiley & Sons, New York-London, Bd. 1: »Chemistry«, 1962; Bd. 2: »Technology«, 1964.

[6] *Benning, C. I.:* »Plastics Foams«, Bd. 2: »Structure Properties and Applications«, Interscience Publishers, John Wiley & Sons, New York, 1969.

[7] *Knipp, U.:* Dissertation at the Institut für Kunststoffverarbeitung, RWTH Aachen, 1972.

[8] *Piechota, H., H. Röhr:* »Integralschaumstoffe«, Carl Hanser Verlag, München, 1975.

[9] *Müller, E.:* »Polyurethane« in: Houben-Weyl: »Methoden der organischen Chemie«, Bd. XIV/2, 4. Auflage, S. 57, Georg Thieme Verlag, Stuttgart, 1963.

[10] *Spielberger, G.:* Liebigs Ann. Chem. 562 (1949) 99.

[11] *Verley, A., F. Bösing:* Ber. dtsch. chem. Ges. 34 (1901) 3354.

[12] *Reiser, W.:* Farbe und Lack 65 (1959) 370.

[13] *Cooper, W., R. W. Pearson, S. Darke:* Ind. Chemist 36 (1960) 121.

[14] *Brecker, L. R.:* Plast. Engin. (1977) 39.

[15] *Ferstanding, L. L., R. A. Scherrer:* J. Amer. Chem. Soc. 81 (1959) 4838.

[16] *McGinn, C. E., R. G. Spaunburgh:* Discourse at a meeting of the American Chemical Society, Atlantic City, Sept. 1956.

[17] *Buist, J. M., H. Budgeon:* »Advances in Polyurethane Technology«, McLaren & Sons, Ltd., London, 1968.

[18] *Ephraim, S. A., A. E. Woodward, R. B. Mesrobian:* J. Amer. Chem. Soc. 80 (1958) 1326.

[19] *Schwanitz, K.:* Dissertation at the Institut für Kunststoffverarbeitung, RWTH Aachen, 1974.

[20] *Schwesig, H.:* Dissertation at the Institut für Kunststoffverarbeitung, RWTH Aachen, 1978.

[21] *Menges, G., H. Schwesig:* Kautsch. Gummi, Kunstst. 32 (1979) 643.

[22] *Kircher, K., R. Pieper:* Kunststoffe 68 (1978) 141.

[23] *Mrotzek, W.:* »Formteilfertigung durch RSG«, Discours at the 10. Kunststofftechnischen Kolloquiums of the IKV in Aachen, vom 12.–14. 3. 1980.

[24] *Cox, H. W., S. A. Jobst:* Discourse at the 37th, Annual technical conference, ANTEC, 1979, USA.

8.2   Polyisocyanurates and Polycarbodiimides

8.2.1   General

In the presence of suitable catalysts, isocyanates are able to react with themselves (see Section 8.1.3). Dimerization will lead to uretdiones; trimerization will result in isocyanurates. Besides formation of cyclic oligomers, condensation reaction is used to form carbodiimide groups.

$R-N=C=O$

Cyclodimerization → Uretdion

Cyclotrimerization → Isocyanurate

Condensation reaction → $R-N=C=N-R$ Carbodiimide

Each of these types of reaction depends on the choice of an appropriate catalyst. The raw material manufacturer uses cyclodimerization to synthesize specific polyisocyanates. Cyclodimerization is not the task of the polymer processor. In various PUR systems, cyclotrimerization to isocyanurate is carried out side by side with the isocyanate alcohol addition reaction. Cyclotrimerization is a reaction which is also carried out in the processing plant.

Condensation to carbodiimide is accompanied by the elimination of carbon dioxide. The polymer processor runs this reaction in order to fabricate polycarbodiimide foam parts; the unavoidably generated gaseous carbon dioxide is used as blowing agent. For the most part, the formation of carbodiimide proceeds by itself, while the formation of isocyanurate occurs simultaneously with the formation of urethane. The resulting polymers contain various molar ratios of urethane and isocyanurate groups.

It is of commercial importance that polyisocyanurates and polycarbodiimides have a higher heat distortion temperature than polyurethanes. Incorporation of isocyanurate structures into PUR will increase its heat distortion temperature. Pure polycarbodiimide foams melt at high temperatures [1].

8.2.2   Mechanism of Reaction and Catalysis of Isocyanurate Formation

Catalysts used for the synthesis of isocyanurate are very similar to those used for the synthesis of PUR. Catalysts needed for the synthesis of PUR are compounds which have a free electron pair (example: tertiary amines); catalysts employed in the synthesis of isocyanurate are more polarized compounds and may have strictly heteropolar bonds [2, 3]. Gradation in the polarization of the catalyst molecule permits control of the ratio of PUR to polyisocyanurate formation. For example, triethylenediamine (DABCO) is a highly active PUR catalyst. It can be used in a slightly altered form as an effective catalyst for the synthesis of polyisocyanurate foams [4].

Table 40 summarizes the catalysts used for trimerization.

Table 40 Catalysts used for trimerization of isocyanate

Group	Name	Formula
Oxide	Lithium oxide Dibutyl-tin(IV) oxide Trialkyl-arsenic(V) oxide	Li_2O $(Bu_2Sn)_2O$ R_3AsO
Alcoholate	Sodium methylate Potassium t-butylate	$NaOCH_3$ $KOC(CH_3)_3$
Amine	Triethylamine Dimethylbenzylamine Triethylendiamine (DABCO)	$N(C_2H_5)_3$ $N(CH_3)_2CH_2-C_6H_5$ $N_2C_6H_{12}$
Carboxylate	Na-Formate Na-Carbonate Na-Benzoate K-Acetate Ca-Acetate	$Na(HCOO)$ Na_2CO_3 $Na(C_6H_5-CO_2)$ $K(CH_3CO_2)$ $Ca(CH_3CO_2)_2$

In addition to one component catalysts, mixtures of various compounds (for example, amines and epoxides; amines and alcohols; amines and peroxides) are also used as catalysts. Two types of catalysts are important: the combination of a tertiary amine with an epoxide, and the alkali salts of weak organic acids [3]. The primary step of the reaction is the addition of one catalyst molecule onto the carbon atom of the isocyanate group to form a catalyst complex (a); to which another isocyanate molecule is added on in the secondary step (b). The newly formed dimer may be stabilized by decomposition of the complex and formation of a uretdione. But it is possible for the dimer to add on another isocyanate molecule and then rearrange to the stable isocyanurate by cleavage of the catalyst molecule (Eq. c) [5].

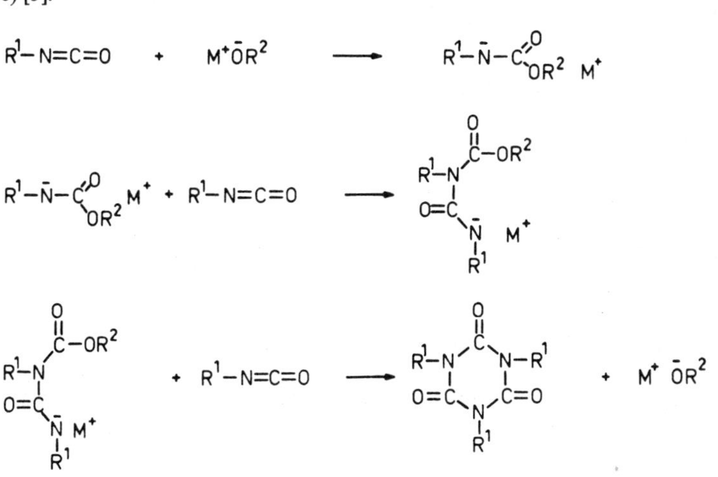

The catalysis by a tertiary amine plus an epoxide is distinguished by a preliminary reaction:

Tertiary amine   Epoxide   Catalyst for formation of isocyanurate

Initially, the tertiary amine acts as catalyst for the synthesis of PUR. The PUR reaction has a head start since the isocyanurate catalyst is not formed before the early stages of the curing process.

8.2.3   Synthesis of Polyisocyanurate Foams [6]

Since pure polyisocyanurate foams are very brittle, cyclotrimerization is used only in conjunction with PUR formation in the polymer processing plant. This process will enrich the PUR system with di- or polyisocyanate; excess isocyanate is converted into a polymer by cyclotrimerization.

In order to produce polyisocyanurate and PUR in one operation, a certain time interval between the formation of urethane and the formation of isocyanurate is needed. Isocyanate, necessary for the formation of urethane, cannot be consumed by a competing reaction if total conversion of the polyols is desired. The use of suitable catalysts will achieve the necessary balance between the two competing reactions.

For the synthesis of polyisocyanurate foams, the same equipment can be used as for the synthesis of PUR foam. Even the systems of operation are very similar. Differences exist in the choice of activators and in the ratio of polyol and polyisocyanate.

Increasing the concentration of the catalyst will result in early trimerization [5a]. The ratio of isocyanurate to urethane formation is dependent on the temperature and the type of catalyst; it is also dependent on the type of isocyanate and polyol component.

The characteristic numbers (CN) used for synthesis of polyurethane polyisocyanurate polymer vary between 110 and 600, and even higher characteristic numbers may be used; however, this is of little interest, since the resulting polymers are very brittle.

8.2.4   Reaction Mechanism of Formation of Carbodiimide

Certain phosphorus compounds such as

Phospholines   Phospholidines   Phospholinoxides   Phospholinsulfides

are suitable catalytic accelerators for the formation of carbodiimides [1]. While autocatalytic formation of carbodiimide proceeds at higher temperatures, catalysts are able to start the reaction at room temperature.

Spontaneous formation of polycarbodiimide takes place when the catalysts mentioned above are mixed with diisocyanates at room temperature; the resulting polymer is a foam, since formation of carbodiimide from isocyanates is a condensation and eliminates carbon dioxide. Without the addition of blowing agents, the resulting foams are expanded by CO_2 only, and if every isocyanate group is converted to carbodiimide, their density depends mostly on the percentage of NCO of the employed polyisocyanate. One hundred grams of MDI with 30% of NCO by weight will yield 15.7 g of carbon dioxide, which equals 8 liters; theoretically, parts will be obtained that have a gross density of 12.5 kg/m³. If diisocyanates are used exclusively, linear chain molecules will be formed.

The catalytically effective phosphorus compounds are very strongly polarized; they combine with the strongly polar isocyanate molecule and form an unstable complex which will break down to carbon dioxide and a new, highly activated intermediate (Eq. a). The intermediate combines quickly with an additional isocyanate group, and a new intermediate is formed; cleavage will produce carbodiimide and restore the catalyst (Eq. b).

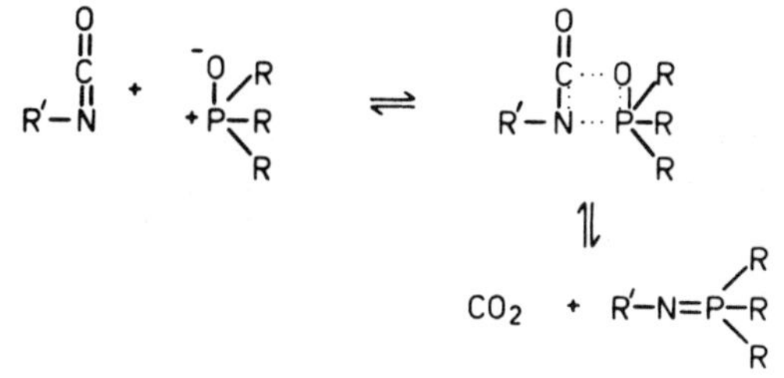

(a)

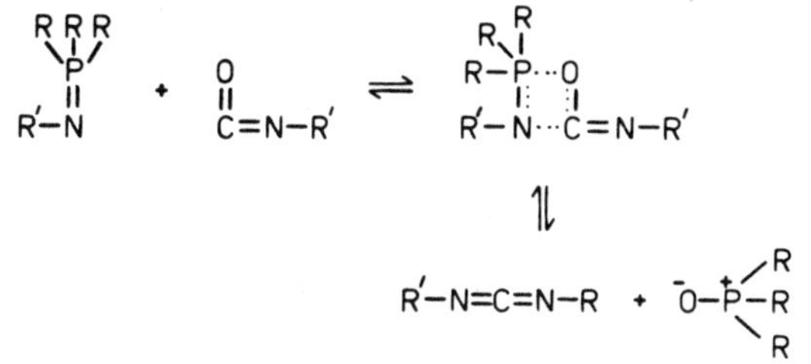

(b)

An induction period is necessary for the formation of isocyanurate; however, catalysts that are used for the formation of polycarbodiimide act spontaneously and are highly effective. Immediately after addition of the catalyst the polymer is formed. This makes it difficult to form a homogeneous blend of catalyst and diisocyanate. Therefore, modified catalysts are being used; these are complex compounds consisting of the phosphorus compound and a ligand. Modifying agents are:

Mono-, di-, or polyalcohols, Proton acids, Metal salts, or Acid chlorides [7].

The complex catalyst (for example, phosphorus compound + HCl) has to dissociate before catalysis can start; this established dissociation makes practical stirring and starting times possible and still sufficiently activates the formation of carbodiimide.

8.2.5   Synthesis of Polycarbodiimide Foams [7]

Polycarbodiimide foams are synthesized without the use of polyols. Isocyanate is the only reactive component. The first step in the synthesis of rigid foam from polycarbodiimide is the blending of isocyanate with an additive (approximately 5% by weight of total resin mixture), which acts as a plasticizer in the finished product. The catalyst (approximately 4–6% by weight of isocyanate) is added shortly before synthesis. The extreme ratio of the mixture (100:5) has to be considered when selecting the equipment. Installations for the synthesis of PUR foam cannot usually achieve this mixing proportion. The term "characteristic number" (CN) is important for the synthesis of PUR foam but does not exist for the synthesis of polycarbodiimide. The ratio of isocyanate to activator determines the rate of reaction and therefore has to be kept constant.

The temperature of isocyanate is approximately 20–25 °C during the synthesis of polycarbodiimide foam. The heat of reaction is low for the formation of polycarbodiimide and for large-scale reactions, the temperature reached in the core is no higher than 70 °C; production of foam pieces of all sizes is therefore possible. A decrease in concentration of catalyst allows foaming at higher temperatures.

During the synthesis of polycarbodiimides the viscosity of the resin increases linearly with time. It is impossible to determine a gel point. Only foam pieces of simple geometric design can be synthesized, since the viscosity increases very rapidly at the beginning of the reaction. The resulting foam is open-celled.

Bibliography to Section 8.2

[1] *Joel, D., G. Behrendt:* Plaste Kautsch. 23 (1976) 162–165.
[2] *Vieweg, R., A. Höchtlen:* Kunststoff-Handbuch, Bd. VII, »Polyurethane«, C. Hanser Verlag, München, 1966.
[3] *Behrendt, G., D. Joel:* »Zur Katalyse von Urethan-Isocyanurat-Polymeren«, Plaste Kautsch. 23 (1976) 177–180.
[4] *Sayigh, A. A. R.:* »Chemistry and Properties of Urethane Foams« in: »Advances in Urethane Science and Technology«, Vol 3, Technomic Publishing Co., Inc., Westport, Conn., USA, 1974.
[5] *Ulrich, H.:* »Unconventional Chemistry of Isocyanates« in: »Urethanes in Elastomers and Coatings«, Technomic Publishing Co., Inc., Westport, Conn., USA, 1973.
[5] a) *Bachara, J. S., R. L. Mascioli:* Discourse at the 7. Intern. Fachtagung für Schaumkunststoffe, May 1977, in Düsseldorf.
[6] *Moos, E. K., D. L. Skinner:* J. Cell. Plast., Part I 12 (1976) 332, Part II 13 (1977) 276, Part III 13 (1977) 399, Part IV 14 (1978) 143, Part V 14 (1978) 208.
[7] *Mann, M.:* Discourse at the 5. Intern. Fachtagung für Schaumkunststoffe, May 1975, in Düsseldorf.

8.3   Epoxy Resins

8.3.1   General

Epoxy resins are prepolymers which are characterized by an epoxy group:

Epoxy group

Before curing, epoxy resins are mixed with a curing agent. Several resin hardener mixtures cure spontaneously at room temperature; other mixtures require higher temperatures for curing. Choosing the proper ratios of epoxy resin to curing agent will result, in the ideal case, in a product without epoxy groups but with newly formed secondary hydroxyl groups. Setting of epoxy resins produces compounds which can no longer be called epoxy resins, since they are not resinous and do not contain epoxy groups. These products called either "cured epoxy resins" or "epoxy resin molding material" [1].

Most epoxy resins contain hydroxyl groups as well as epoxy groups. The chemical structure of the remaining molecule may vary widely; this explains.the multitude of types of epoxy resins. Usually, relatively large amounts of curing agent are used, and therefore the chemical structure of the curing agent will influence the composition and properties of the resulting epoxy resin polymers.

The curing of epoxy resins is a polyaddition reaction. To a small extent, epoxy resins are cured by polymerization of the epoxy group. Detailed descriptions of the synthesis, processing, and properties of epoxy resins and their polymers are given in [2-4]. The epoxy group is less reactive than the isocyanate group. Epoxy resins require longer curing times than the synthesis of PUR. Also, epoxy curing differs from PUR synthesis by the wide variation in the ratio of epoxy resin to curing agent. While the polymer reaction of PUR using a mixture of polyisocyanate and polyol cannot be stopped intentionally by lowering the temperature, this opportunity exists for the curing of epoxy resins with certain types of curing agents. Therefore, it is quite easy to synthesize and store sheet molding compound (SMC) made from epoxy resins. A process for manufacturing SMC from PUR has been developed only recently [5].

Characteristic properties of epoxy resins are:
- Very low shrinkage (caused by reaction), which makes epoxy resins suitable as encapsulating compounds
- Ease of processing by reaction casting at room temperature and atmospheric pressure without elimination of volatile compounds
- Good mechanical properties and inertness to corrosive materials
- good electrical properties
- Exceptionally large variability of properties of resin and polymer

Only a small part of the total production of epoxy resins is used for reaction casting. Most of it is used in other applications. In 1980 in the United States, only 8% of the total production was used for the manufacturing of parts and 42% as resins for paints and coating material.

The use of certain types of curing agents requires higher temperatures for the cure which makes these resin-hardener systems stable at room temperature (one-component resins). In order to effect curing, the stable resins are heated and poured into preheated molds, or the molding compounds are heated by microwaves.

Two-component resins are more important than one-component resins from a tonnage standpoint. In the two component system, the epoxy resin and curing agent are handled separately and are mixed shortly before processing. Both components may be liquid, highly viscous, or solid compounds. Heating will transform solid or highly viscous products into more easily workable materials.

Occasionally, the viscosity of the raw material is adjusted by so called reactive diluents. These are compounds with low viscosity that participate in the curing reaction of epoxy resins.

8.3.2   The Chemistry of the Epoxy Group

The epoxy group is a cyclic ether that consists of a three-member ring. The bond angles are greatly strained and the ring is under stress. By reacting with various other compounds, the ring will open and form a new and less strained molecule. The epoxy group is polarized by the different electronegativities of carbon and oxygen. Oxygen has a negative partial charge and carbon is partially positively charged. The epoxy group is present mostly in the form of glycidyl ether:

$$R - O - \overset{3}{C}H_2 - \overset{2}{C}H - \overset{1}{C}H_2 \qquad \text{Glycidyl ether}$$

Because of the one-sided substitution, each of the carbon atoms of the epoxy ring has a different partial charge. Carbon atom number-1 carries the strongest positive partial charge. This charge distribution is the cause for the true reaction, the addition of compounds with removable hydrogen atoms (acidic hydrogen atoms) onto the epoxy ring:

$$R_1 - CH - CH_2 \; + \; H - R_2 \longrightarrow R_1 - \underset{OH}{CH} - CH_2 - R_2$$

The epoxy oxygen atom remains with carbon atom 2 of the epoxy resin; but addition of the other component takes place in the end position (carbon atom 1).

The following compounds (in decreasing reactivity) are the most widely used reactants for the curing process of epoxy resins:

- Amines
- Carboxylic acids
- Alcohols

Amines with at least one available free hydrogen atom (on the nitrogen atom) react very quickly with epoxy resins at relatively low temperatures. "Amine hardeners" are compounds which have at least two hydrogen atoms on the nitrogen atom. In practice, compounds with two amino groups are used as curing agents. Amino hardeners react very quickly, even at low temperatures; therefore, they are employed in cold-temperature curing. For this reason, a mixture of epoxy resin and amino hardener is of low stability unless, at very low temperatures, the amine is converted into a salt complex or enclosed in a molecular sieve, this process will make the amine available only at higher temperatures.

<table>
<tr><td colspan="3" align="center">Reactions of the epoxy group</td></tr>
</table>

$R_1-CH-CH_2$ (epoxy) $\quad+\quad$	R_2-NH_2 primary Amine	$\xrightarrow[\text{temp.}]{\text{Room}}$ $\quad R_1-CH-CH_2-NH-R_2$ $\qquad\qquad\;\; \overset{|}{OH}$ secondary Amine
$R_1-CH-CH_2$ (epoxy) $\quad+\quad$	$\overset{R_2}{\underset{R_3}{\diagdown}}NH$ sec. Amine	$\xrightarrow[\text{temp.}]{\text{Room}}$ $\quad R_1-CH-CH_2-N\overset{\diagup R_2}{\diagdown R_3}$ $\qquad\qquad\;\; \overset{|}{OH}$ tertiary Amino group
$R_1-CH-CH_2$ (epoxy) $\quad+\quad$	R_4-COOH Carboxylic acid	$\xrightarrow[\text{temp.}]{\text{Low}}$ $\quad R_1-CH-CH_2-O-\overset{\overset{O}{\|}}{C}-R_4$ $\qquad\qquad\;\; \overset{|}{OH}$ Ester group
$R_1-CH-CH_2$ (epoxy) $\quad+\quad$	R_5-OH Alcohol	$\xrightarrow[\text{temp.}]{\text{High}}$ $\quad R_1-CH-CH_2-O-R_5$ $\qquad\qquad\;\; \overset{|}{OH}$ Ether group
$R_1-CH-CH_2$ (epoxy) $\quad+\quad$	R_6-SH Mercaptan	$\longrightarrow$ $\quad R_1-CH-CH_2-S-R_6$ $\qquad\qquad\;\; \overset{|}{OH}$ Thioether group
$R_1-CH-CH_2$ (epoxy) $\quad+\quad n\; R_1-CH-CH_2$ (epoxy)	$\xrightarrow{\text{Catalyst}}$	$\bar{O}-CH-CH_2\!\left[\!O-CH-CH_2\!\right]_{n-1}\!O-CH-\overset{+}{C}H_2$ $\qquad\;\; \underset{R_1}{|} \qquad\quad \underset{R_1}{|} \qquad\qquad\quad \underset{R_1}{|}$

Strong polarization of the O-H bond of carboxylic acids ensures fast reaction between epoxy groups and carboxylic acids. Carboxylic groups act as activators and accelerate the reactions of the epoxy group. Addition of the acidic hydrogen atom onto the ether oxygen will increase the positive charge of the epoxy ring and make the molecule more reactive to addition of nucleophilic reactants:

$$-CH-CH_2 \;+\; H^+ \;\longrightarrow\; -CH-\overset{+}{C}H_2$$

Polarization of the hydroxyl group is minimal, and curing by alcohol based hardeners requires temperatures of 100–200 °C. The low reactivity of the hydroxyl group allows the coexistence of epoxy and hydroxyl group in the same molecule.

At lower temperatures, reaction does not occur. Even at higher temperatures the rate of reaction is so slow that this reaction cannot be used as the sole curing reaction.

Polymerization of the epoxy group is of great importance. The ability to polymerize is given by the polarization of the ring structure.

$$-CH-CH_2 \rightleftharpoons -CH-\overset{+}{C}H_2$$

$$\overset{+}{C}H_2-CH + \overset{+}{C}H_2-CH \longrightarrow \overset{+}{C}H_2-CH-O-CH_2-CH \longrightarrow$$

$$\overset{+}{C}H_2-CH-O-CH_2-CH + n\ CH_2-CH \longrightarrow \overset{+}{C}H_2-CH\left[O-CH_2-CH\right]_n O-CH_2-CH$$

In addition to the reaction between the epoxy group, reaction of the secondary hydroxyl group and the acid anhydrides is of particular importance:

$$R_1-C\overset{O}{\diagdown}\\ R_2-C\overset{O}{\diagup} O\ +\ HO-R_3 \longrightarrow R_1-\overset{O}{\overset{\|}{C}}-OH\ +\ R_2-\overset{O}{\overset{\|}{C}}-O-R_3$$

The reaction between acid anhydride and alcohol will yield a free carboxyl group and an ester group. While the acid anhydride does not react with the epoxy molecule, it will react with the hydroxyl group of the epoxy resin and form a carboxylic group which in turn is able to react with the epoxy ring.

8.3.3   Classification of Epoxy Resins

Usually epoxy resins are classified into three groups:
- Epoxy resins based on polyphenols (epoxy base resins)
- Aliphatic and cycloaliphatic epoxy resins
- Nitrogen-containing epoxy resins

This classification is based on historical reasons only. The oldest and, in volume of production, the most important epoxy resins are the ones based on polyphenols. Aliphatic and cycloaliphatic epoxy resins came on the market only recently and are produced in smaller quantities. Reactive amines are used for the synthesis of epoxy resins containing nitrogen, but the effect of amines on the reactions of the epoxy group is very similar to that of phenols or olefins.

Epoxy resins can be classified according to the technical process:
- Cold setting resins (room temperature)
- Thermosetting resins ($< 100\,°C$)
- High temperature thermosetting resins ($< 100\,°C$)

Usually, curing conditions do not depend on the chemical structure of the polyepoxy compound but on the type of hardener and catalyst.

Usually, room-temperature curing systems use amines as hardeners; high-temperature curing systems employ anhydrides which allow cross-linking not only between hydroxyl group and anhydride but also between newly formed carboxyl groups and epoxy groups. Addition of catalysts will lower the curing temperature, and these resins can be classified as moderate-temperature curing resins.

8.3.4   Epoxy Resins Based on Polyphenols

Reaction products of bisphenol-A with epichlorohydrin represent the largest tonnage amount of all epoxy resins. The synthesis of epoxy resins based on bisphenol-A and epichlorohydrin will be described briefly since this will easily clarify the differences among the various types of resins.

At high temperatures, both hydroxyl groups of bisphenol-A react with epichlorohydrin to form glycerol monochlorohydrin ether (Eq. a) which is then converted to bisphenol-A diglycidyl ether by reacting with sodium hydroxide (Eq. b).

In the presence of a large excess of epichlorohydrin the reaction will yield bisphenol-A diglycidyl ether exclusively. This compound is the least complicated epoxy resin and has the lowest viscosity of all bisphenol-A resins.

Use of a small excess of epichlorohydrin leads to further reaction of bisphenol-A diglycidyl ether with bisphenol-A. Chain propagation takes place (Eq. c), resulting in compounds with high molecular weight.

Further reaction between bisphenol-A and epoxy groups will yield high molecular weight compounds which, under ideal conditions, contain an epoxy group at each end of the chain and, with increasing chain length, a growing number of secondary OH groups.

Commercially available epoxy resins based on bisphenol-A vary essentially only in their chain length.

The percentage by weight of epoxy groups decreases with increasing molecular weight. Viscosity of the resin increases with increasing chain length. Compounds with higher molecular weight are solids at room temperature. Because of side reactions, the high molecular weight epoxy resins will contain fewer epoxy groups than theoretically calculated.

Table 41   Polyphenols as starting materials for the synthesis of epoxy resins

Name	Formula
4,4 Dihydroxydiphenyl-2,2-propane (bisphenol-A)	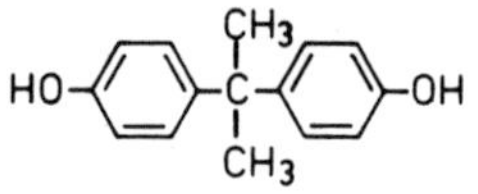
Tetrabromobisphenol-A	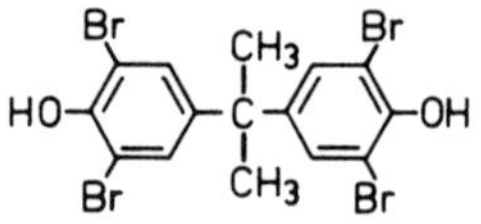
Dihydroxydiphenylmethane (bisphenol-F)	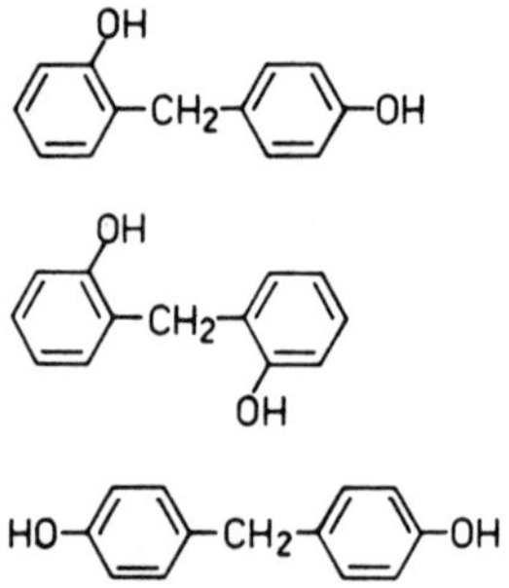
Novolak resins (intermediates for producing phenoplasts	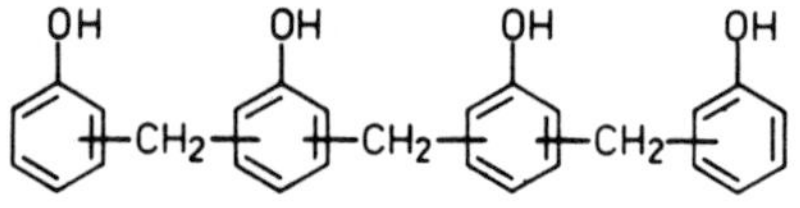

In addition to bisphenol-A resins, the literature describes a large number of resins which are based on polyphenols (Table 41). Only those products which can be processed economically are of commercial importance. Raw materials for bisphenol-A are phenol and propylene; both are moderate in price and easily available. Following bisphenol-A in commercial importance are the lower molecular weight condensation products of phenol and formaldehyde, especially bisphenol-F and novolak. Tetrabromobisphenol-A is used for custom made products, for instance, for the manufacture of extremely flame-resistant thermosetting epoxy resins.

8.3.5 Epoxy Resins Based on Cycloaliphatic Compounds and Amines

Cycloaliphatic epoxy resins (see Table 42) contain one or more cycloaliphatic rings. The absence of aromatic character in these resins results in polymers with different properties than those made from standard resins. Special properties of the thermosetting cycloaliphatic resins are greater weather stability and greater resistance to (electric) surface currents (tracking and arcing).

Cycloaliphatic resins can be synthesized by two different chemical methods. One method uses cyclic olefins which are converted into epoxy resins by "epoxidation". Another method uses cycloaliphatic polycarboxylic acids which are reacted with epichlorohydrin and then converted to epoxy compounds by reaction with bases.

Similar methods are employed for the synthesis of epoxy resins based on amines. Here, synthesis of the necessary amines is the first step; then the amines are reacted with epichlorohydrin and converted into epoxy resins by treatment with bases.

Table 42 a Epoxy resins based on cycloaliphatics

Name	Formula
Vinylcyclohexene-dioxide	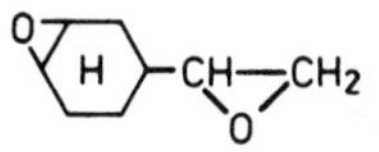
Bis-(2,3-epoxycyclo-pentyl)-ether	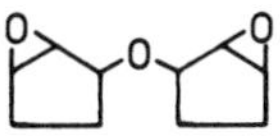
3,4-Epoxycyclohexyl-methyl-3,4-epoxycyclo-hexanecarboxylate	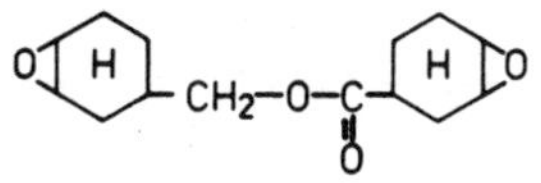
Bis-(3,4-epoxy-6-methyl-cyclohexyl-methyl)-adipate	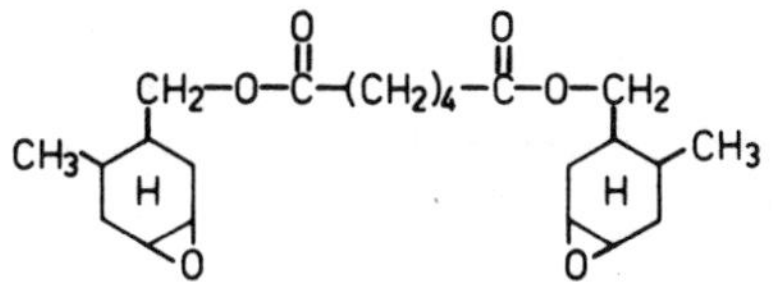
Hexahydrophthalic acid diglycidylester	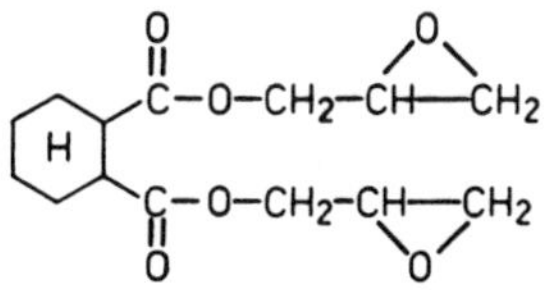
Trimethylpropane-tri-(2,3-epoxypropylhexa-hydrophthalate)	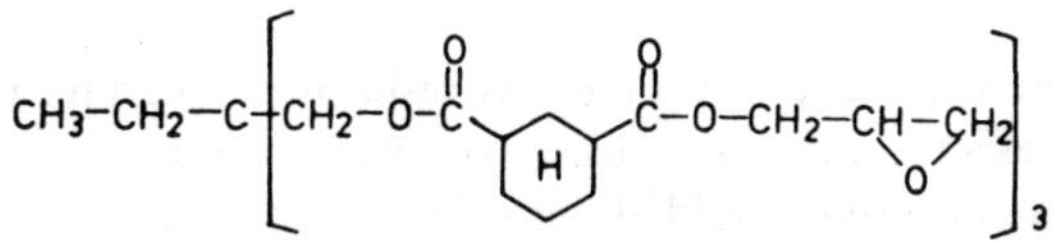

(continued on next page)

Table 42b   Epoxy resins based on cycloaliphatics and amines

Name	Formula
Tetrahydrophthalic acid diglycidyl ester	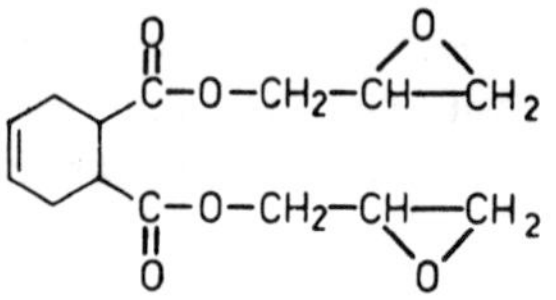
Triglycidyl-isocyanurate	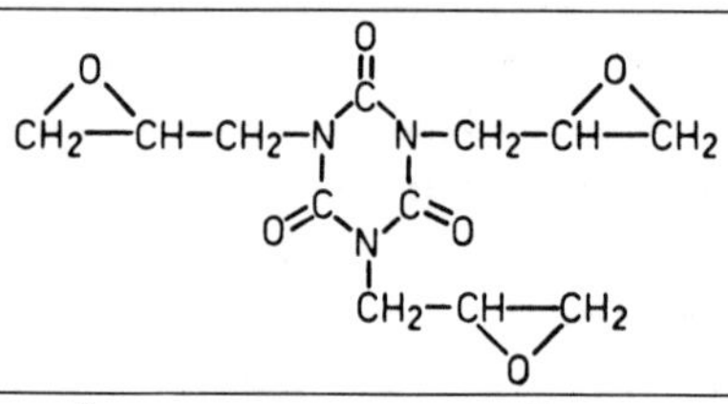
N,N-Diglycidylaniline	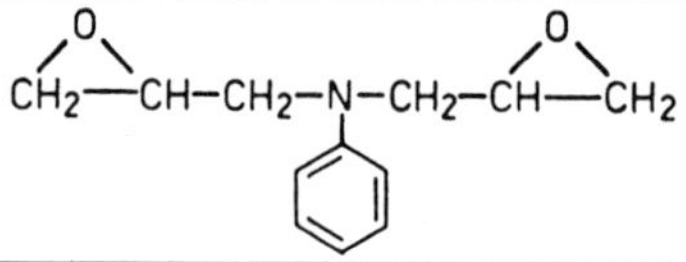
N,N,N′-N′-Tetraglycidyl-4,4′-diaminodiphenylmethane	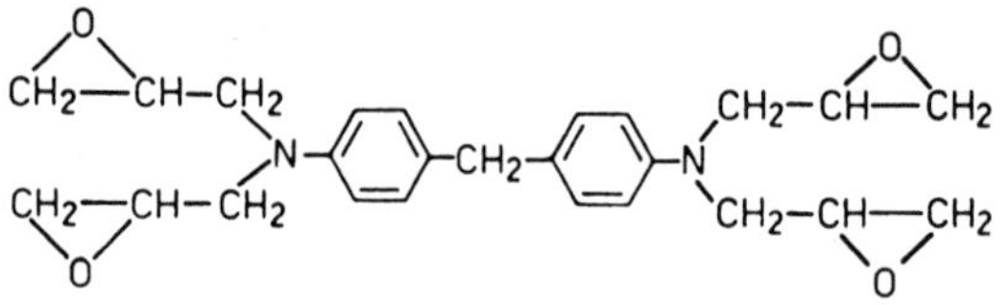
2,4-Diglycidyl-5,5-dimethylhydantoin	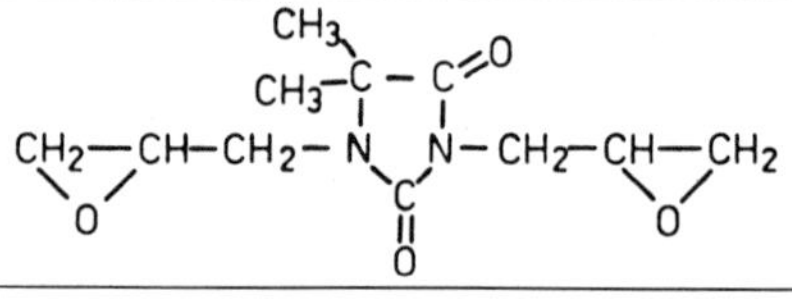
2,4-Di(2-glycidylpropyl)-5,5-dimethylhydantoin	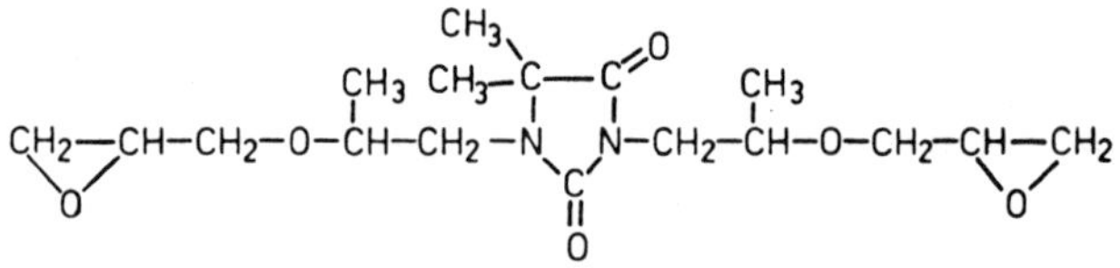

8.3.6   Reactive Diluents

The viscosity of several epoxy resins is too high for most purposes, and elevated temperatures and additives are used to reduce the viscosity. The use of solvents is not suitable in the manufacture of molded parts. The use of low-viscosity compounds containing epoxy groups is excellent for lowering the viscosity. The terms "reactive thinner" and "epoxy resin" overlap each other. Low viscosity epoxy resins are sometimes used as reactive thinners for more viscous epoxy resins or for epoxy resins which are solid at room temperature.

Table 43   Reactive thinners for epoxy resins

Name	Formula	Viscosity at 25 °C (m.Pa.s)	Boiling pt. (°C/mbar)
Allylglycidylether	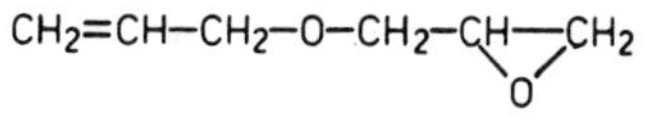 $CH_2=CH-CH_2-O-CH_2-CH{-}CH_2$ (epoxide)	1,09	63-63,5/40
Butylglycidylether	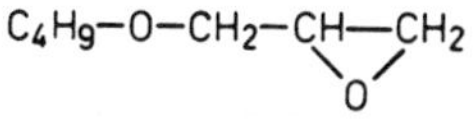 $C_4H_9-O-CH_2-CH{-}CH_2$ (epoxide)	1,34	80,5/50
Methacrylic acid glycidyl ester	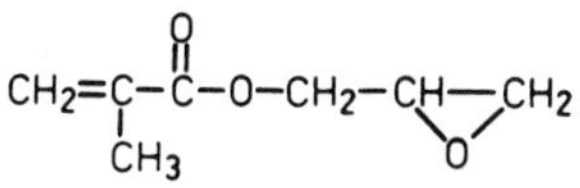 $CH_2=C-C-O-CH_2-CH{-}CH_2$ with CH_3 and O	2,24	70,5-71/8
Tetrahydro furfuryl glycidyl ether	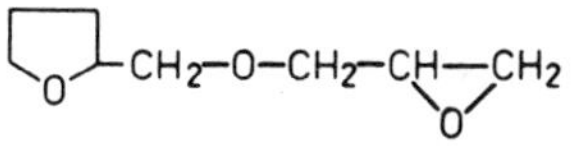 $-CH_2-O-CH_2-CH{-}CH_2$ (with furfuryl and epoxide)	3,77	92-93,5/5,3
Cresyl glycidyl ether	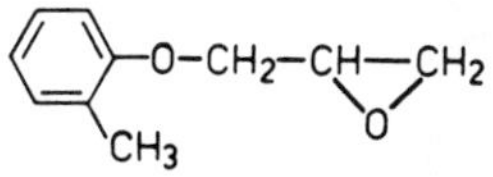 $-O-CH_2-CH{-}CH_2$ (with CH_3, aromatic and epoxide)	5	260
Phenyl glycidyl ether	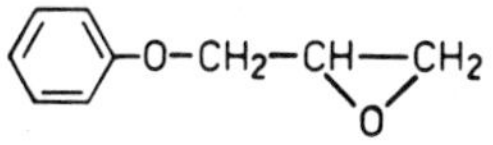 $-O-CH_2-CH{-}CH_2$ (aromatic, epoxide)	5,67	106,5/6,7
Benzoic acid glycidyl ester	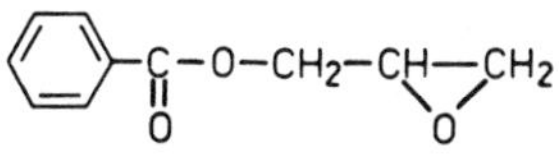 $-C-O-CH_2-CH{-}CH_2$ (aromatic, O, epoxide)	9,30	112,5-113/2,7
1,4 Butanediol diglycidyl ether	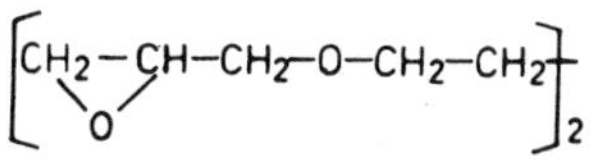 $[CH_2-CH-CH_2-O-CH_2-CH_2]_2$ (epoxide)	16,6	240

Most of the reactive diluents have only one epoxy group which is grafted to the epoxy resin; after this grafting process is performed, they lose their ability to further lower the viscosity. In addition to that, most reactive thinners are used as modifying agents for the cured resins. For instance, reactive thinners are used to incorporate segments into the resin which will make it more elastic. Glycidyl ethers containing long chains of saturated carbon atoms (C_8 to C_{24}-alcohols) are particularly suitable for this purpose.

The literature describes a large number of reactive diluents [7]. Table 43 lists examples of reactive diluents which are not used simultaneously as epoxy resins. Some of the reactive diluents have a low boiling point and are therefore suitable only for use with resins that set at room temperatures.

A special type of reactive diluent are polymerizable diluents such as methyl methacrylate and styrene; they are converted to vinyl polymers by free radical polymerization during the curing of epoxy resins [8, 9].

8.3.7   Curing of Epoxy Resins with Anhydrides

Table 44 lists the anhydrides suitable as curing agents. The low price of phthalic anhydride makes it the most important agent. Its high melting point is a disadvantage. 3,6-dimethyl tetrahydrophthalic anhydride and methylnadic acid anhydride are liquids at room temperature, but their price is higher than that of phthalic anhydride. The highly chlorinated

Table 44   Anhydride curing agents for epoxy resins

Acid anhydride	Formula	Mol wt. (theoretical)	Boiling pt. [°C]
Phthalic anhydride		148,1	130...131
Tetrahydrophthalic anhydride		152,2	100...101
Hexahydrophthalic anhydride		154,2	35... 37
Methyl nadic anhydride		178,2	liquid
Hexachloroendomethylene tetrahydro phthalic anhydride (HET anhydride)		370,8	239...240
Pyromellithic acid dianhydride		218,1	284...286
Maleic anhydride		98,1	52... 53
Dodecylsuccinic anhydride		266,4	liquid

HET anhydride is used for the synthesis of flame-resistant products. Dodecylsuccinic anhydride is liquid at room temperature and yields elastic polymers as a result of its long-chain paraffin group.

Epoxy resins with no hydroxyl groups cannot be cured by carboxylic acid anhydrides. If anhydrides are to be used for the curing process the resin has to contain secondary hydroxyl groups or small amounts of components with exchangeable hydrogen atoms such as alcohols or amines. During the initial step, the compound containing the hydrogen atom attacks the anhydride; this can be seen as a reaction between the anhydride and the hydroxyl groups of the resins (see Eq. a). During this reaction the curing agent combines with the resin to form half esters, because the anhydrides of carboxylic acids which are used in this process are always cyclic anhydrides of dicarboxylic acids. This formation of an epoxy resin containing a carboxyl group is an important step of the reaction. The newly formed carboxyl group reacts quickly with other epoxy groups and forms an additional ester and a (free) hydroxyl group (see Eq. b).

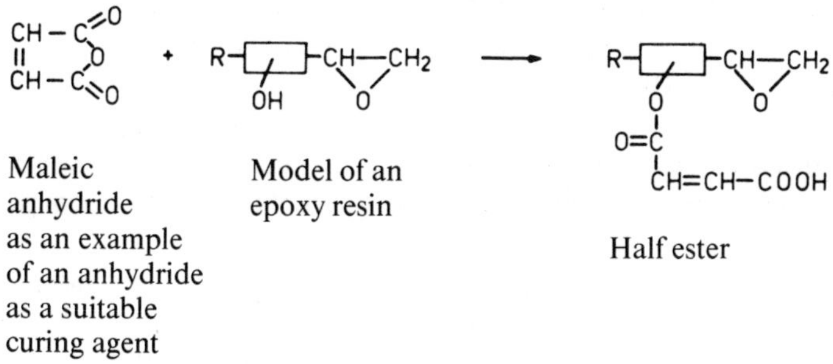

Maleic anhydride as an example of an anhydride as a suitable curing agent

Model of an epoxy resin

Half ester

(a)

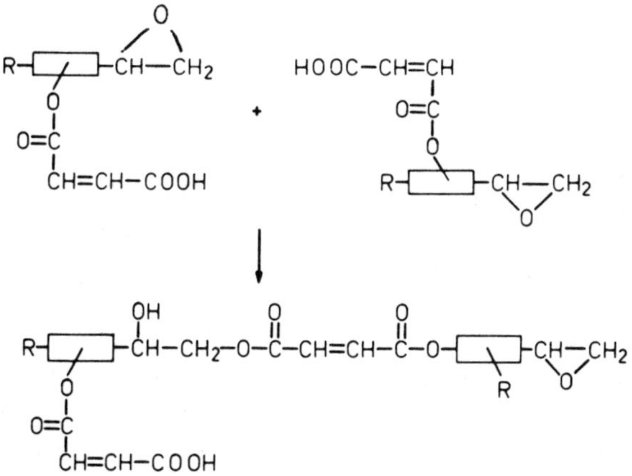

(b)

Each of these cross-linking steps, created by the reaction with an anhydride, produces a secondary hydroxyl group which is able to function as a reactant for further reactions. Initiation of the curing reaction of epoxy resins which are free of hydroxyl groups needs only small amounts of compounds containing exchangeable hydrogen atoms. The concentration of these additives does not change during the curing process. Initiation of the curing reaction is made possible by a foreign substance.

Polyhydroxyl compounds cannot be used as curing agents for epoxy resins; reactions between hydroxyl groups and epoxy groups require high temperatures and even then proceed very slowly. During the curing with anhydrides, the reaction between hydroxyl and epoxy groups occurs as a side reaction, especially at higher temperatures. Hydroxyl groups, which are part of the resin, are in competition with carboxyl groups resulting, in the formation of ethers:

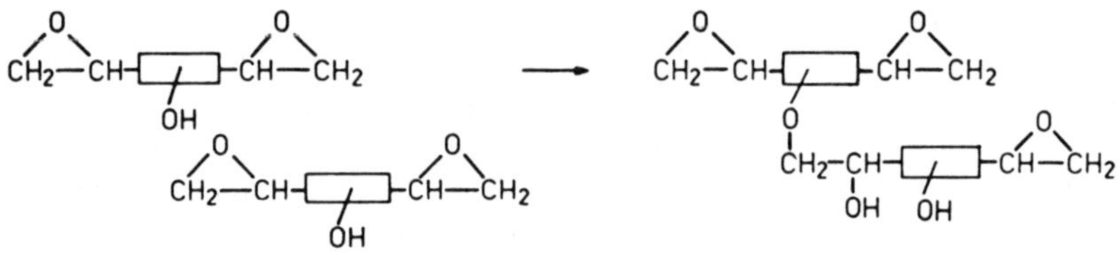

Acidic compounds act as catalysts for the formation of esters as well as for the formation of ethers, during which carboxyl groups, formed during the reaction, are able to function as acidic catalysts. Formation of ethers is unimportant, since the catalysts accelerate the formation of esters to a much greater extent than the formation of ethers.

Mixtures of epoxy resins and pure anhydride (curing agent) are very stable and have a long shelf life, since carboxyl groups which act as catalyst are formed only during the reaction and initiation of this reaction requires temperatures higher than 100 °C. Anhydrides of technical grade are not very pure; they contain various amounts of free carboxyl groups and therefore add a certain amount of catalyst to the resin.

Water contains a specific type of hydroxyl group. Water is available in the resin either as a remainder from the synthesis or as a subsequent impurity. It reacts with the anhydride in the same way as the hydroxyl groups of the resin, resulting in the formation of a dicarboxylic acid which participates with both carboxyl groups of the acid in the cross-linking reaction.

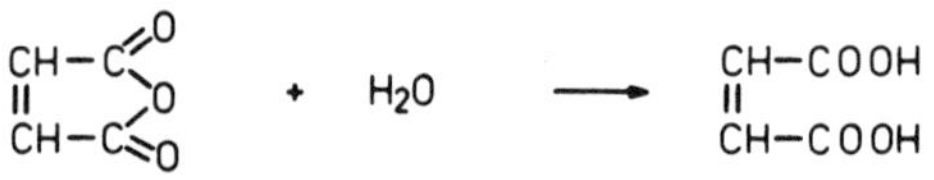

An additional reaction, if unimportant in scope, is the reaction between free carboxyl groups and hydroxyl groups which forms in esters; as a result of this condensation water is liberated. This does not cause any trouble as long as the water is consumed by the reaction with anhydride.

The progress of the curing of epoxy resin can be monitored with infrared spectroscopy, which determines the amounts of epoxy anhydride and carboxyl groups still available in the system as well as the amounts of newly formed hydroxyl and ester groups (Figure 31). Measuring the dielectric dissipation factor and the specific inductive capacity (dielectric constant) is another method to monitor the progress of the curing, because both of these values are directly proportional to the concentration of functional groups [10].

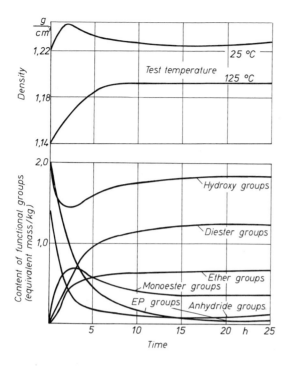

Figure 31
Change in density and
number of functional
groups during curing of
EP resins with phthalic
acid anhydride [11].
Resin: Araldit B,
Curing agent: 31.4 parts
of PSA to 100 parts resin.
Temperature: 125 °C

8.3.8 Curing of Epoxy Resins with Amines

Table 45 gives a summary of amines customarily used as curing agents. Low molecular weight amines have the disadvantage of having a high vapor pressure. Preliminary reactions of amines with small amounts of epoxy resins or conversions of the amines into acid amides will reduce their vapor pressure.

The curing reaction with amines is described by the following equations:

R—NH$_2$ + CH$_2$—CH—[ ]—CH—CH$_2$ $\longrightarrow$ NH—CH$_2$—CH—[ ]—CH—CH$_2$

Amine curing agent   Epoxy resin

CH$_2$—CH—[ ]—CH—CH$_2$ + NH—CH$_2$—CH—[ ]—CH—CH$_2$

$\downarrow$

CH$_2$—CH—[ ]—CH—CH$_2$—N—CH$_2$—CH—[ ]—CH—CH$_2$

Secondary hydroxyl groups are formed also during the curing with amines. The first step of the reaction is formation of a secondary amine which still contains an exchangeable hydrogen atom and is therefore still reactive. Monoamines act as bifunctional compounds (toward epoxy resins) and are able to convert diepoxy resins into high molecular weight macromolecules through chain propagation. Diamines are tetrafunctional compounds and

Table 45   Reactive amines as curing agents

Name	Formula	Boiling pt. [°C]	Required quantity[1]
Ethylenediamine	$H_2N-CH_2-CH_2-NH_2$	11	6...8
Diethylene-triamine	$H_2N-(CH_2)_2-NH-(CH_2)_2-NH_2$	−39	9...11
Triethylene-tetramine	$H_2N-[(CH_2)_2-NH]_2-(CH_2)_2-NH_2$	−35	10...13
m-Phenylene-diamine	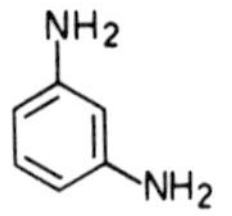 (NH_2 benzene ring NH_2)	63	13...15
4,4'-Diaminodi-phenylmethane	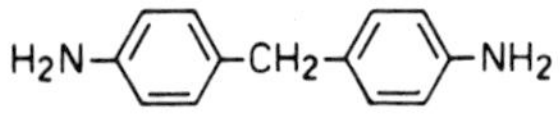 $H_2N-\bigcirc-CH_2-\bigcirc-NH_2$	93	26...28
4,4'-Diaminodi-phenylsulfone	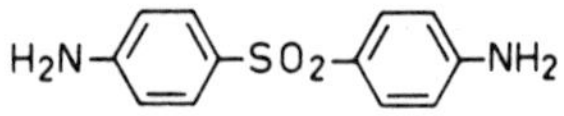 $H_2N-\bigcirc-SO_2-\bigcirc-NH_2$	177	33...38
Polyaminoamide	$HO-[C(=O)-R-C(=O)-NH-R'-NH]_n H$	viscous	50...100

[1] Amount of amine necessary for curing refers to 100 parts of a low molecular weight dian resin which has an equivalent mass of epoxy of 190 g [1].

yield highly cross-linked products. At certain quantitative ratios, all amino groups are converted into tertiary amines. Anhydrides react as curing agents at higher temperatures only, but mixtures of epoxy resins and amines cure at room temperature.

Tertiary amines do not have any free hydrogen atoms and are therefore unable to react with epoxy resins, but they are able to act as catalysts to reactions of epoxy groups with carboxyl, hydroxyl, and amino groups.

8.3.9   Curing of Epoxy Resins by Catalytic Polymerization

In addition to accelerating the setting of epoxy resins with anhydrides and amines, tertiary amines activate polymerization of these resins. Various tertiary amines are used as catalysts for the setting reaction and as activators for the setting with anhydrides. A list of polymerization activators is given in Table 46. Basic as well as acidic compounds are able to act as activators; boron trifluoride complexes are acidic compounds which are frequently used as catalysts.

Addition of the catalyst onto the epoxy ring results in ring scission and formation of an amphoteric ion (Eq. a). Continuous addition of the epoxy group to the negatively charged oxygen results in polymerization of the epoxy resin (Eq. b).

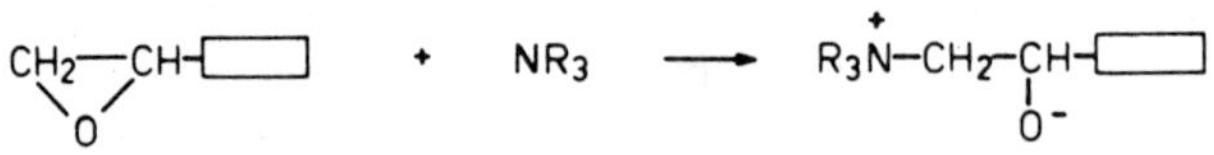

$$CH_2-CH-\square \quad + \quad NR_3 \quad \longrightarrow \quad R_3\overset{+}{N}-CH_2-CH-\square \atop O^- \qquad \text{(a)}$$

Table 46   Catalytically effective curing agents for epoxy resins

Name	Formula	Melting pt. (theoretical)	Required quantity[1]
N,N'-Dimethyl-benzylamine	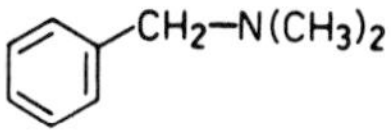 C_6H_5—CH$_2$—N(CH$_3$)$_2$	135,2	5...10
α-Methyl N,N-dimethyl benzylamine	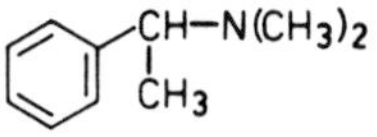 C_6H_5—CH—N(CH$_3$)$_2$ / CH$_3$	149,2	5...10
o-(N,N-dimethyl-aminomethyl)-phenol	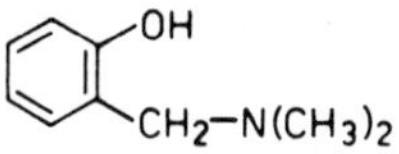 C_6H_4(OH)—CH$_2$—N(CH$_3$)$_2$	151,2	6...10
2,4,6-Tris-(N,N-dimethyl-aminomethyl)-phenol	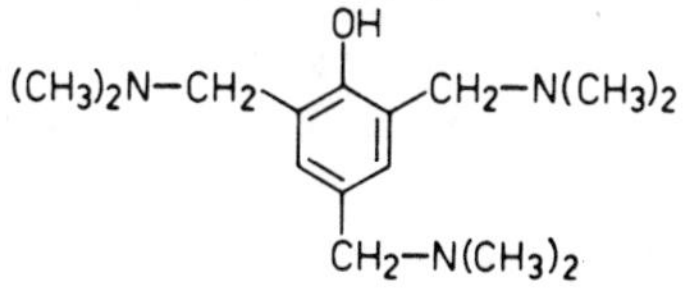 (CH$_3$)$_2$N—CH$_2$—[C$_6$H$_2$(OH)]—CH$_2$—N(CH$_3$)$_2$ / CH$_2$—N(CH$_3$)$_2$	265,4	6...10
Piperidine	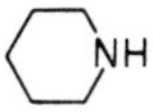 C_5H_{10}NH	85,2	6
Benzylamine-BF$_3$ complex	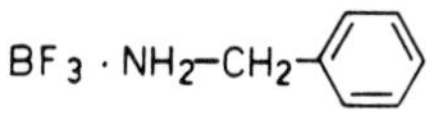 BF$_3$ · NH$_2$—CH$_2$—C_6H_5	175	1...5
N-Methylaniline-BF$_3$ complex	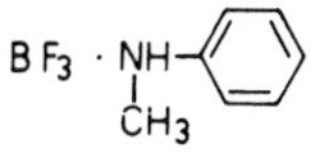 BF$_3$ · N(CH$_3$)H—C_6H_5	175	1...5

[1] related to 100 parts resin [1]

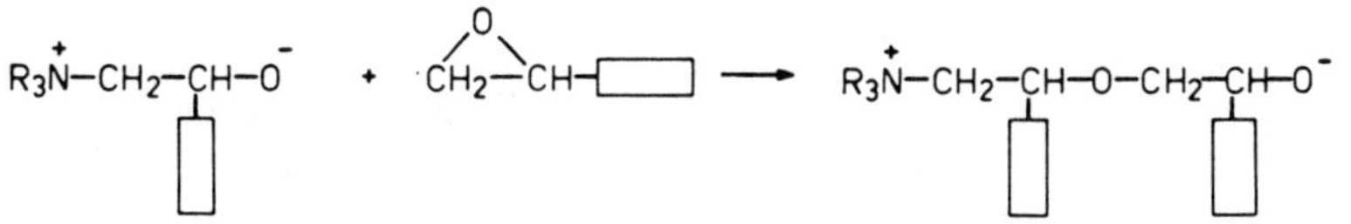

$$R_3\overset{+}{N}-CH_2-CH-\overset{-}{O} \; + \; CH_2-CH-\square \longrightarrow R_3\overset{+}{N}-CH_2-CH-O-CH_2-CH-\overset{-}{O} \qquad \text{(b)}$$

Elimination of the activator R$_3$N group and proton transfer results in termination of the reaction (Eq. c).

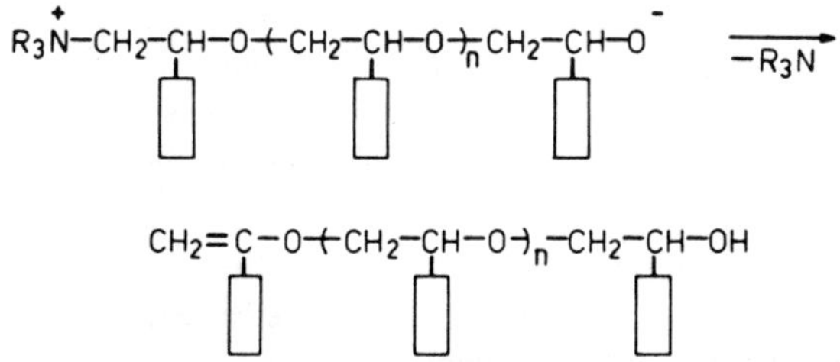

$$R_3\overset{+}{N}-CH_2-CH-O+CH_2-CH-O+_nCH_2-CH-\overset{-}{O} \quad \xrightarrow{-R_3N} \qquad \text{(c)}$$

$$CH_2=C-O+CH_2-CH-O+_n-CH_2-CH-OH$$

8.3.10   Ratio of Resin to Curing Agent

Theoretically, the most favorable ratio of resin to hardener is equimolar amounts of reactive groups in the resin mixture. However, this theoretically calculated ratio is not the most advantageous, since side reactions occur, and formation of cross-linked structures will result in incomplete conversion of the reactants. The most favorable ratio must be determined empirically.

Elevated temperatures increase the mobility of chains and promote conversion of functional groups. Products which are synthesized at lower temperatures are tempered to improve product qualities. Even this step does not result in total conversion of all functional groups. A higher degree of conversion can be reached by using an excess of curing agent. But this excess of curing agent can have negative effects on the properties of the polymer.

The criterion to define the optimum mixing ratios is the quality of the cured molding compounds. The determination of the optimum value has to be carried out separately for each individual resin hardener combination, and it has to consider the effects of various curing conditions. Usually, the raw material manufacturer specifies the ratio. A deviation of up to 10% in the ratio will alter the qualities of the polymer very little; this is a definite advantage of the epoxy resins.

Curing of the epoxy resins by polymerization does not follow stoichiometric rules. The hardener acts as a catalyst and is used in very small amounts only. An exact determination of the quantitative ratio is possible only by experimentation. The optimum concentration of catalyst is measured by the optimum properties of the molding compound and is between 1 and 10% by weight.

8.3.11   Heat of Reaction and Shrinkage of Epoxy Resins during Curing

Every resin has a specific heat of reaction which is released during curing. Therefore, a numerical value can be given only for a specific resin. The heat of reaction for 1 mole of epoxy group converted is 92–109 kJ according to [12]; this is higher than the molar heat of reaction for polymerization of vinyl compounds.

The equivalent mass of epoxy resin varies between very low values (below 100 g per mole of epoxy groups) and very high values (above 1000 g per mole of epoxy groups for high molecular weight epoxy resins). The value for the heat of reaction varies accordingly. The setting of a resin containing a high number of epoxy groups with amines and additional catalysis will very quickly result in releasing the heat of reaction, and consequently the temperature will rise from room temperature to above 200 C. If the setting reaction occurs slowly, it is easier to discharge the heat of reaction and thus, avoid the high increase in bulk temperature.

The total amount of shrinkage which occurs during setting is the sum of shrinkage caused by the reaction and thermal effects. The amount of shrinkage caused by reaction is proportional to the conversion of functional groups; it depends on the specific type of resin. Also, shrinkage is greatly affected by the type of curing agent used (see Figure 32). It is assumed that curves 3 and 4 (Figure 32) do not show completed reactions and that the indicated low values for shrinkage are not the final values.

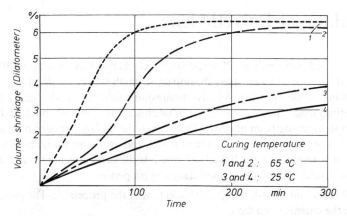

Figure 32 Volume contraction during curing of an epoxy resin (Epon 824, equivalent weight 185–195) [1]
Parts of curing agent per 100 parts of resin: 1, diethylamine propylamine (8 parts); 2, N phenylenediamine (14.5 parts); 3, diethylenetriamine (8 parts); 4, Triethylenetetramine (10 parts).

Although shrinkage of epoxy resins is very low, its value of 5–6% by volume is similar to the values for PUR and UP resins. In comparison to UP resins, gelation of epoxy resins occurs relatively late; therefore, high flow compensates for most of the curing shrinkage.

Epoxy resins with an equivalent mass of epoxy of approximately 1000 exhibit only one-fifth of the shrinkage shown in Figure 32, since concentration of epoxy groups is lowered by the same degree. These resins exhibit shrinkage of only 1% by volume; it is possible to lower this value even further by addition of fillers and careful manipulation of flowing properties prior to reaching the gel point. Also, lower shrinkage takes place if hardeners are used which have a high molecular weight and contain a small number of reactive groups.

Bibliography to Section 8.3

[1] *Jahn, H.:* »Epoxidharze«, VEB Deutscher Verl. f. Grundstoffindustrie, Leipzig, 1969.
[2] *Vieweg, R., M. Reiher, H. Scheurlen:* Kunststoffhandbuch, Bd. XI, »Polyacetale, Epoxidharze, fluorhaltige Polymerisate, Silicone usw.«, C. Hanser Verlag, München, 1971.
[3] *Lee, H., K. Neville:* »Handbook of Epoxy Resins«. McGraw-Hill-Book-Company, New York, 1967.
[4] *Paquin, A. M.:* »Epoxidverbindungen und Epoxidharze«, Springer Verlag, Berlin-Göttingen-Heidelberg, 1958.
[5] *N. N.:* »Baypreg« – Messe-Information K'79, Bayer AG.
[6] *N. N.:* Mod. Plast. Intern. 11 (1981) 29.
[7] *Pilny, M., J. Mleziva:* »Reaktive Verdünner für Epoxidharze«, Kunststoffe 67 (1977) 783.
[8] *Kubens, R.:* Kunststoffe 64 (1974) 666.
[9] *Kubens, R.:* Kunststoffe 69 (1979) 455.
[10] *Menges, G., A. Hille, W. Thiele:* Kunststoffe 69 (1979) 467.
[11] *Fisch, W., W. Hofmann:* Makromol. Chem. 44/46 (1961) 8.
[12] *Klute, C. H., W. Viemann:* J. Appl. Polym. Sci. 5 (1961) 86.

9 Cross-linking of Polyethylene

A sparsely cross-linked polyethylene (PE) (approximately 5 cross-linking sites per 1000 carbon atoms) exhibits improved creep strength, impact strength (in the cold), and resistance to stress cracking while showing slightly diminishing hardness and rigidity [1]. Stretched films (sheeting) made from cross-linked PE (shrink films) generate much higher shrinking forces than those made from noncross-linked material [2]. The amount of cross-linking in PE is very important to the manufacture of foams and insulators for high-voltage cables. The synthesis of foams from cross-linked polymer is technically advantageous, since the resulting products exhibit exceptionally good damping properties.

In practically all cases, the cross-linking reaction is part of the processing. The processor is responsible for the chemical reaction.

Several processes are available which can be used to cross-link PE:
- Cross-linking by UV light in combination with photolytically cleavable free radical generators
- Cross-linking induced by thermally cleavable free radical generators
- Silane cross-linking
- UHF cross-linking

9.1 Overall Chemical Reactions

Two chemically different methods of cross-linking PE can be distinguished according to the reactions involved: (a) dimerization of the polymer radical is the essential step of the cross-linking reaction and (b) the polymer is cross-linked by a reaction between reactive, functional groups:

(a) Cross-linkage by dimerization of the radical:

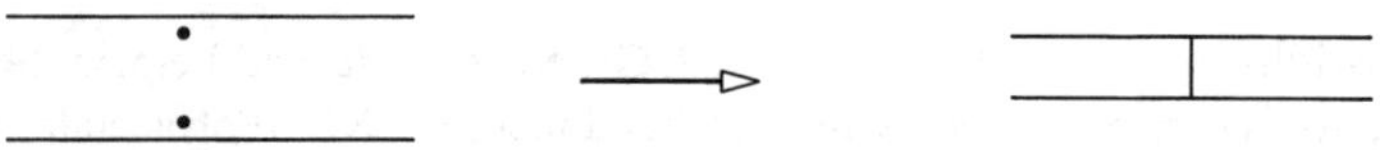

(b) Cross-linkage by reaction between functional group (example):

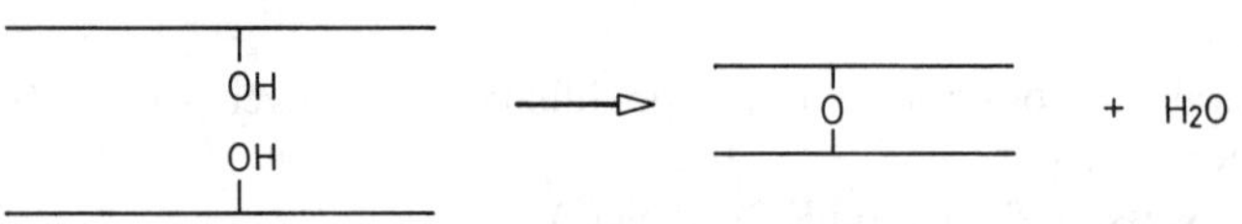

Processes which make use of the dimerization of polymer radicals are commercially very important. Various methods are available to generate the necessary free polymer radicals, while the dimerization reaction is the same for all processes. Polyethylene radicals are generated by cleavage of a C-H bond:

$$-CH_2-CH_2-CH_2-CH_2- \longrightarrow -CH_2-\overset{\cdot}{C}H-CH_2-CH_2- + H\cdot$$

Energy may be used to cleave the C-H bond, which has a bond energy of 364 kJ/mole (for secondary hydrogen atoms); the energy of electron beams is sufficient to serve as the source of energy. No additional auxiliary components are needed for electron beam cross-linking; energy alone is sufficient for the reaction.

Other methods for cross-linking make use of radical transfer from primary radicals to PE for the generation of the polymer radicals needed for dimerization. A preliminary reaction generates primary radicals which act as auxiliary agents. This reaction determines the rate of the cross-linking reaction; primary radicals themselves are incorporated into the polymer only by a side reaction if at all.

Cross-linking by free radicals consists of three steps (unless electron beams are used):

(a) Generation of primary radicals:

$$\text{Radical generator} \xrightarrow{+\text{Energy}} R\cdot + \dots$$
$$R\cdot = \text{Primary radical}$$

(b) Transfer of radicals to the polymer molecule, followed by abstraction of hydrogen atoms:

$$R\bullet \; + \; -CH_2-CH_2-CH_2-CH_2- \; \longrightarrow \; R-H \; + \; -CH_2-\underset{\bullet}{C}H-CH_2-CH_2-$$

(c) Dimerization of radicals:

$$\begin{array}{l} -CH_2-\underset{\bullet}{C}H-CH_2-CH_2- \\ -CH_2-\underset{\bullet}{C}H-CH_2-CH_2- \end{array} \longrightarrow \begin{array}{l} -CH_2-\underset{|}{C}H-CH_2-CH_2- \\ -CH_2-CH-CH_2-CH_2- \end{array}$$

According to this mechanism, the polymer chain does not incorporate products which are formed either by decomposition of radical generators or by transfer of radicals from the primary radical. These compounds remain as by-products in the cross-linked polymer and may affect the properties of the final product.

The chain radicals are active components which have a strong tendency to form a paired electron covalent bond, thus returning to a lower state of energy. Dimerization represents the desired formation of covalent bonds and is the essential step in the reaction, but other possible reactions do exist. Chain radicals can lose their activity either by chain cleavage or by formation of an olefin.

Formation of olefin:

$$-CH_2-\underset{\bullet}{C}H-CH_2-CH_2- \; \longrightarrow \; -CH_2-CH=CH-CH_2- + H\bullet$$

Chain splitting:

$$-CH_2-\underset{\bullet}{C}H-CH_2-CH_2-CH_2- \; \longrightarrow \; -CH_2-CH=CH_2 \; + \; \bullet CH_2-CH_2-$$

Dimerization through incorporation of the primary radical (R) represents an additional side reaction:

$$-CH_2-\underset{\bullet}{C}H-CH_2-CH_2- \; \xrightarrow{+R\bullet} \; -CH_2-\underset{\underset{R}{|}}{C}H-CH_2-CH_2-$$

Any of these reactions is possible and all may occur at the same time, depending on the conditions. Chain cleavage is promoted by high temperatures. In addition to that, chain cleavage is promoted by an increasing number of tertiary hydrogen atoms in the chain. The shifting of the radical along the polyethylene chain is a side reaction which is very important to the course of the cross-linking reaction [3]. The free hydrogen shift is only possible as long as the bond energy is the same for all C-H bonds.

Hydrogen shift; shifting of the radical site:

$$-CH_2-\overset{\bullet}{C}H-CH_2-CH_2- \quad \longrightarrow \quad -\overset{\bullet}{C}H-CH_2-CH_2-CH_2-$$

The cross-linking by radicals differs from cross-linking by polycondensation, which is a process involving chemical reactions of the side chains. Examples of the latter are the cross-linking of polyethylene sulfochloride and the silane cross-linking in which condensation of Si-OH groups and subsequent cross-linking to Si-O-Si groups occurs. Cross-linking of polyethylene sulfochloride, together with other cross-linking processes, is described in Section 10.6.

9.2   Cross-linking by Electron Beams

The cross-linking of PE by electron beams has been known for a long time and is discussed extensively in [4–8]. During irradiation of PE with electrons having a voltage of 2 MeV, the electrons will penetrate the polymer and interact with the resin. As a rule of thumb, electrons with a beam voltage of 2 MeV will penetrate up to 1 cm of the PE polymer [8]; energy density decreases with the thickness of the layer, and therefore this process is used mainly for thin products such as films, and the cross-linking of molded parts with thick walls will result in variable cross-linking density. The main applications of the cross-linking process using electron beams are in the manufacture of shrink hoses, shrink films, shrinkable insulating parts, and cross-linking of insulating cables and foams [9]. The degree of interaction between the electron beam and the polymer depends on the local energy density and also in the polymer's state of aggregation and the surrounding atmosphere. Variables in PE which can be controlled experimentally are: formation of a three-dimensional network

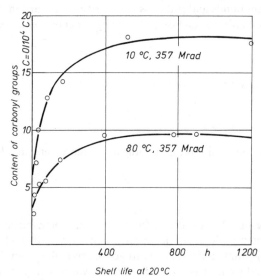

Figure 33
Formation of carbonyl groups at the surface of high-density PE films after irradiation by electron beams in vacuum as a function of subsequent exposure to air [14]

of bonds, chain branching, and formation of low molecular weight products such as waste gas which, in the case of strictly linear PE, consists of pure hydrogen and low molecular weight hydrocarbons [10–12].

At room temperature, polymer radicals are quite stable and lose their radical activity rather slowly. Any radical sites remaining in the finished product will undergo spontaneous reaction with oxygen that has infiltrated by diffusion and form carboxyl groups [13] (see Figure 33). Also, the chain radicals are able to initiate polymerization of a monomer which has been applied to the PE surface. Parts made from irradiated PE can be coated with other polymers by this method [15].

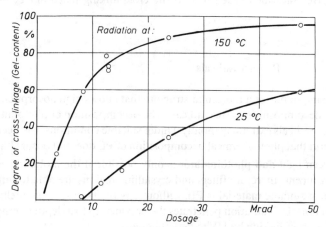

Figure 34   Gel concentration in linear PE after electron irradiation under nitrogen [16].

The percentage of cross-linking (expressed as percentage of gel present) increases with the radiation dose (see Figure 34). The blending of PE with polymerizable additives prior to irradiation will increase the efficiency of radiation or lower the dosage necessary to achieve a certain percentage of cross-linking [17]. Addition of monomer increases the yield of radicals by suppressing other irradiation-induced reactions and by increasing chain mobility, the degree of radical shift, and the number of "hot" hydrogen atoms [17, 18]. The incorporation of monofunctional monomers does not produce an increase in cross-linking density, while the incorporation of monomers with at least two reactive double bonds does. The literature describes the use of acrylic acid [17, 19, 20], diallyl phthalate, allyl methacrylate [20], trially cyanurate, and other triazines containing unsaturated substituents [21, 22], and compounds containing acetylene groups [23–25]. Degradation of PE by irradiation results in molecules containing several unsaturated bonds; these molecules are available for further cross-linking [26].

Cross-linking by electron beams has the advantage of being usable on cold objects. Cross-linking of pure PE, without any additives, is an additional advantage of the above-described process.

Radiation cross-linking of polymers in the melt will result in products with different physical properties than cross-linking at lower temperatures [27]. It has been shown that only amorphous PE will be cross-linked by irradiation at room temperature; in the crystalline regions of the polymer, radicals are generated, but cross-linking does not occur [28]. Therefore, it is more difficult to cross-link HDPE than LDPE by the electron beam method.

9.3   Cross-linking by UV light

Polyethylene absorbs UV light, and the absorbed energy is sufficient to photolytically cleave C-H bonds. This process will generate the free radicals needed to cross-link the polymer, similar to the electron beam process [29]. The process in too slow and is accompanied by too many side reactions to be of commercial value. Mixtures of PE and radical generators which can be cleaved photolytically make it possible to use cross-linking by UV light in a commercial process. The principal action of the UV light is to decompose the photoinitiators and to generate the free radicals needed to start the cross-linking reaction (see Section 6.2.1):

$$R - R \xrightarrow{\quad h \cdot \nu \quad} 2 \; R \cdot$$

Photoinitiator   Primary radicals

Photoinitiators are compounds of a molecular structure that allows high absorption of UV light, which results in decomposition to free radicals. The decomposition of photoinitiators generates primary free radicals extremely rapidly. Under suitable conditions; the necessary reaction time is so short that photochemical decomposition of PE does not occur [30].

An interaction between light and photoinitiator is only possible if the molding compound is sufficiently translucent; therefore fillers and crystalline regions are able to influence the reaction. The most suitable material is PE, without any fillers; above the melting point of the crystalline regions. UV radiation penetrates the polymer up to a depth of only a few millimeters; this limits cross-linking by UV to very thin parts.

Ketones such as benzophenone [31, 32] and benzil dimethyl ketal [33] are suitable photoinitiators for cross-linking of PE. In addition to initiators which are solids at room temperature, the use of gaseous compounds such as chlorinated hydrocarbons and disulfur dichloride [34], phosgene, thionyl chloride, sulfuryl chloride, and the corresponding bromine compounds [35] and carbon oxysulfide is described in the literature [36]. These volatile photoinitiators are easily used with polymers that have already been cast into a film; the film is treated with volatile photoinitiators and then cross-linked by exposure to UV radiation. Cross-linking by electron radiation is enhanced by addition of polymerizable monomer which takes an active part in the cross-linking process by forming cross-linked bonds. The same method can be employed for cross-linking by UV radiation [37].

High temperatures will accelerate cross-linking, since PE itself will react faster photochemically and the rate of decomposition of the photoinitiator will increase (see Figure 35).

But since high temperature promotes chain cleavage, cross-linking at higher temperatures is accompanied by chemical decomposition. An increase in radiation intensity will increase the number of decomposition reactions (Figure 36).

The most suitable application for cross-linking by UV radiation is in the cross-linking of PE films. Broader use of cross-linking by UV radiation has not yet developed.

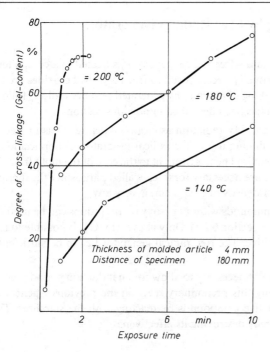

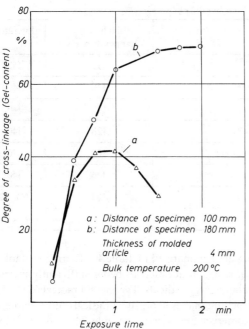

Figure 35
Cross-linking of LDPE with 2% (by weight) benzyl dimethyl ketal; degree of cross-linking (gel content) as a function of the distance from the radiation source (high-pressure mercury lamp, 100 W/cm). [38]

Figure 36
Cross-linking of LDPE with 2% by weight benzyl dimethyl ketal; degree of cross-linking (gel content) as a function of the distance from the radiation source (high-pressure mercury lamp, 100 W/cm) [38].

9.4   Thermal Cross-linking by Free Radical Generators

In this process a mixture of PE and a free radical generator is heated to a certain temperature at which the additive is thermally decomposed to free radicals. Peroxides are the preferred additives [39]. The cross-linking of polymer by generated primary radicals follows the same reaction mechanism which has been described in previous sections.

While the use of electron rays and UV radiation as cross-linking agents is limited to parts with thin walls, thermal cross-linking using free radical generators is applicable for parts of any thickness. Usually, the required heat is brought to the core of the mold from the outside; very long cross-linking times are necessary for thick walled parts, since the thermal conductivity for PE (~0.5 W/mK for density of 0.95 g/cm) is very low.

A few peroxides and several compounds without peroxy or azo groups can be used for the thermal cross-linking process (see Section 6.2.1). Only above 115 °C is it possible to mix LDPE homogeneously with an additive. The free radical generator has to be sufficiently stable at this temperature.

The presence of stabilizers makes it necessary to allow for an induction period preceding the cross-linking reaction. During this preliminary reaction the previously generated free radicals are converted into inactive compounds by reacting with the stabilizer. This reaction consumes the stabilizer thereby destroying its effectiveness.

Table 47   Peroxides which can be used as cross-linking agents for PE

Peroxide	10 h half life Temperature (°C)	Literature
Dicumylperoxide	115	[40]
α,α-Bis(t-butylperoxy)-diisopropyl-benzene	116	[40]
2,5-Dimethyl-2,5-di-(t-butylperoxy)-hexane	119	[40]
2,5-Dimethyl-2,5-di(t-butylperoxy)-hexin-(3)	128	[40]
4,4-Bis(t-butylperoxy)-valeric acid-n-butyl ester	109	[40]
Di-t-butylperoxide	126	[40]
t-Butylperbenzoate	107	[41, 42]
3-Phenyl-3-(t-butylperoxy)-phthalide		[43]

Peroxides suitable for cross-linking are summarized in Table 47. Thermally stable hydroperoxides are unsuitable for cross-linking. In addition to being sufficiently thermally stable, the peroxides have to generate high energy radicals. The methyl radical (CH_3) has a particularly high energy. Increased branching makes short-lived radicals more stable, and radicals which can be stabilized by mesomerism are even more stable. The order of stability is as follows:

Radical transfer is responsible for the formation of free radical structures in the polymer molecule. The methyl radical is the most reactive radical and can react with all hydrocarbons. In the linear PE system, the elimination of hydrogen will produce secondary radical sites which are more reactive than primary radicals in tertiary position, therefore, the latter radicals are not very effective cross-linking agents.

Elimination of hydrogen converts tertiary hydrogen atoms of polypropylene and branched PE to tertiary radical chains with low reactivity. Tertiary radical sites are not very reactive and are not converted easily into more reactive secondary radicals. The shifting of the radical site along branched chains is hindered, and dimerization of chain radicals becomes more difficult. Linear HDPE yields the highest number of radicals [27], which means that for HDPE a given concentration of primary radicals will yield the highest number of cross-linking sites.

The cross-linking reaction is dependent not only on the degree of cross-linking but also on the number of olefin double bonds and carbonyl groups [27]. Olefin double bonds promote cross-linking. The mechanism of this cross-linking reaction is similar to that of polybutadiene and polyisoprene; both of these polymers are easily cross-linked by using radicals as cross-linking agents. Carbonyl groups form the initial stage of an oxidative degradation; particularly, they are present in polymers which have been exposed to oxygen either at high temperature or during UV radiation. The cross-linking process using free radicals as cross-linking agents will lead to further decomposition of the polymer without cross-linking at the sites of carbonyl groups. According to [27], PE which has been exposed to air for a long period of time can be cross-linked only to 63% (percent of gel), while unexposed PE cross-linked under similar conditions contains 93% of gel.

A very important and effective peroxide is dicumyl peroxide. Its thermal decomposition results in several reaction products [44, 45]:

(a)

Acetophenone
(b)

Cumene
(c)

α-Methylstyrene
(d)

$$\cdot CH_3 \; + \; PE \; \longrightarrow \; CH_4 \; + \; \cdot PE$$

Methane

The only desirable reaction is the formation of PE radicals (Eq. c and e). Acetophenone is responsible for the intense odor, and methane for the formation of microcavities. 98% of the gas which is generated during reactions of dicumyl peroxide is methane [46].

Decomposition of the peroxide, that is, generation of primary free radicals, is the slowest reaction and therefore the rate-determining step of the three-step cross-linking reaction of PE using peroxides as cross-linking agents. Since thermal decomposition of peroxide proceeds according to the following equation,

$$v = k \cdot [c] \quad \text{with} \quad k = k_o \cdot e^{-\dfrac{E_a}{RT}}$$

v   = rate
k   = rate constant
[c] = concentration of peroxide
E   = activation energy
R   = universal gas constant
T   = absolute temperature

the cross-linking of PE is, in first approximation, dependent on the type of peroxide (k and E) and the temperature (T). The reaction is first-order and depends also on the chemical nature of the polymer to be cross-linked [47, 48]. For LDPE, the value for the order of reaction (depending on the MFI value) is 0.90–0.99; for HDPE, 1.06 and for copolymers of ethylene, between 0.88 and 1.22 (ethylene-butylacrylate-acrylic acid-copolymer, 0.88; ethylene-vinyl-acetate-copolymer, 1.22).

The value of the activation energy for various cross-linking reactions stays practically the same if the same peroxide is used. This illustrates that the decomposition of peroxide is the rate-determining step. The cross-linking reaction is exothermic [49].

It has been shown that the order of reaction is dependent on the temperature and increases at higher temperatures. This is not the case for linear PE except at higher temperatures [47]. It is believed that branching and the promotion of side reactions contribute to that effect.

The methods decribed for the cross-linking by electron beam and by UV light to modify the cross-linking process by concurrent use of polyfunctional, unsaturated polymerizable monomers are the same for the thermally induced free radical cross-linking process [50–52].

Thermal cross-linking by peroxides has the advantage of causing less degradation of polymer than irradiation cross-linking [27]. This advantage can be realized only if high temperatures are avoided. The fact that shearing not only generates heat but may also result in mechanical chain degradation has to be taken in consideration.

A multitude of commercial processes are employed for cross-linking with thermally induced free radicals. These methods differ only in the way in which the necessary heat for the decomposition of peroxide is supplied to the PE-peroxide mixture. A summary is given in [53]. Heat is supplied by heat transfer from the hot gases to PE [54] or by treatment with hot inert liquids, salt solutions, or liquid metals and by treating the mixture with superheated steam or by shear action [55]. Shearing heat occurs between two rotating discs in narrow pipes (channels) [56] or during compression of granular PE-peroxide mixtures [57]. Another method uses molds with heated metal inserts; the part is heated from the inside by thermal conduction [58]. Heating of a mixture consisting of PE, peroxide, and carbon black in an ultrahigh-frequency alternating field represents a totally different method, and will be described in detail in connection with the cross-linking by UHF.

Other methods use IR radiation. Energy absorption increases with increasing thickness of material, which makes this method applicable for only thin films and sheets, which are easily heated by IR radiation. IR radiation is used especially for frothing and simultaneous cross-linking of PE; during this process the expanding agents and peroxides themselves may undergo photolytic decomposition [59, 60].

Since the processor demands short processing, great temperature differences between the surface of the part and its core are unavoidable at the beginning of the cross-linking in parts with thick walls. With this wide range in temperature, cross-linking will start on the surface of the part and then proceed to the core as heat penetrates the mold. Very high surface temperatures will lead to degradation of polymer and subsequent damage of the molding compound.

9.5   Silane Cross-linking

Cross-linking of PE can be achieved by polycondensation of previously incorporated silanol groups (Si-OH) (61–64):

$$-CH_2-CH_2-CH-CH_2-CH_2-$$
$$RO-Si-OR$$
$$O$$
$$H$$
$$OH$$
$$RO-Si-OR$$
$$-CH_2-CH_2-CH-CH_2-CH_2-$$

$$\downarrow -H_2O, \; Cross\text{-}linking$$

$$-CH_2-CH_2-CH-CH_2-CH_2-$$
$$RO-Si-OR$$
$$O$$
$$RO-Si-OR$$
$$-CH_2-CH_2-CH-CH_2-CH_2-$$

The silanol groups are not stable and react spontaneously with themselves without addition of any additives. Therefore, PE modified with silanol groups can not be stored nor can it be processed according to the usual processing methods; its manufacture has to occur simultaneously with the processing.

Overall, silane cross-linking involves a very complicated chemical process. Starting material is a standard PE. The first step of the reaction is the grafting of PE with an alkoxy silane.

$$-\overset{|}{\underset{|}{Si}}-O-R \qquad \text{Silane Alkoxy Group (R = organic radical)}$$

Example: Trimethoxyvinyl silane can be used as the alkoxysilane for silane cross-linking.

$$CH_2=CH-\overset{\overset{OCH_3}{|}}{\underset{\underset{OCH_3}{|}}{Si}}-OCH_3 \qquad \text{Trimethoxyvinyl silane}$$

This silicon compound is sensitive to water and has to be handled carefully. Mixing the PE with peroxide while vinyl silane and at the same time heating the mixture to the decomposition temperature of peroxide will result in grafting the silicon compound onto the macromolecule. It is presumed that primarily a free PE radical is formed (Eq. 2) to which the vinyl compound is added (Eq. 3), followed by the newly formed free radical losing its activity through dimerization or transfer of hydrogen atoms and the formation of a stable compound.

$$R-O-O-R \qquad \xrightarrow{\text{Heat}} \qquad 2\ R-O\bullet \tag{1}$$

Peroxide   Primary radical

$$R-O\bullet + -CH_2-CH_2-CH_2-CH_2-CH_2- \longrightarrow R-OH + -CH_2-CH_2-\overset{\bullet}{CH}-CH_2-CH_2 \tag{2}$$

chain radical

$$-CH_2-CH_2-\overset{\bullet}{CH}-CH_2-CH_2- + CH_2=CH-\overset{\overset{CH_3}{\overset{|}{O}}}{\underset{\underset{CH_3}{\overset{|}{O}}}{Si}}-OCH_3 \longrightarrow \begin{matrix} -CH_2-CH_2-CH-CH_2-CH_2- \\ | \\ CH_2 \\ | \\ \bullet CH \\ | \\ CH_3O-\overset{|}{\underset{\underset{OCH_3}{|}}{Si}}-OCH_3 \end{matrix} \tag{3}$$

The polymer molecule containing the trimethoxysilane groups can be stored if it is kept in moistureproof containers and the polymer is commercially available in appropiate packaging. According to the method described above, a side reaction of free-radical-induced cross-linking of PE would be possible, theoretically. Extended cross-linking would make further processing of the intermediate impossible, but it has been observed that in practice these cross-linking reactions are of little consequence.

The modified PE which has been prepared in a separate preliminary reaction is mixed with a master batch containing a catalyst and is converted into a molded article. At an appropriate time, water is added and methoxysilane groups are quickly hydrolyzed to silanol groups with the help of the catalyst; methanol is eliminated. Overall the cross-linking process represents a condensation reaction. Only hydrolysis is able to generate the active group essential for cross-linking (Eq. 4):

$$\begin{matrix} -CH_2-CH_2-CH-CH_2-CH_2- \\ | \\ CH_2 \\ | \\ CH-R \\ | \\ CH_3O-\overset{|}{\underset{\underset{OCH_3}{|}}{Si}}-OCH_3 \end{matrix} \qquad \xrightarrow[\substack{/+ \text{ Catalyst } / \\ -3\ HOCH_3}]{+\ 3\ H_2O} \qquad \begin{matrix} -CH_2-CH_2-CH-CH_2-CH_2- \\ | \\ CH_2 \\ | \\ CH-R \\ | \\ HO-\overset{|}{\underset{\underset{OH}{|}}{Si}}-OH \end{matrix} \tag{4}$$

Silane cross-linking seems to be chemically very complicated but is relatively easy in practice. Usually, modified PE is purchased, mixed with a master batch containing a catalyst and processed. The resulting molded part is not yet cross-linked. Subsequent diffusion of water initiates the cross-linking reaction. Of advantage here is the high rate of diffusion of water into PE. Placing the parts in water at higher temperatures (approximately 90 °C) will finish the cross-linking reaction in an economically acceptable time span. Also, it is possible to couple the grafting reaction with the processing procedure; silane and peroxide are fed together with the esterification catalyst into the feed hopper of a screw extruder [65, 66].

Other compounds such as y-methacryloxy propyl trimethoxysilane [67] or vinyltri-ethoxysilane [68] are also suitable silane components.

$$CH_2=C-\overset{\overset{O}{\|}}{\underset{\underset{CH_3}{|}}{C}}-O-CH_2-CH_2-CH_2-\overset{\overset{O-CH_3}{|}}{\underset{\underset{O-CH_3}{|}}{Si}}-OCH_3$$

Approximately 2% by weight of silane components is needed; the amount of peroxide required for grafting is approximately 0.15%. These peroxides are the same as the ones used for other PE cross-linking processes. Only very small amounts (0.05%) of catalyst are needed for siloxane hydrolysis. Suitable catalysts are:

- Dibutyltin dilaurate [65]
- Tetrabutyl titanate [67]

In silane cross-linking, every alkoxysilane group has the same chance of becoming a cross-linking site. Every silicon atom is therefore a cross-linking site for several PE chains [69].

Silane cross-linking is suitable for cross-linking of PE parts without exceeding the softening point of the molding compound during the reaction. A very promising application seems to be the production of extruded pipes made from cross-linked PE; the pipes are extruded and subsequently cross-linked by treatment with water. The patent literature describes methods for the production of PE foams made from PE modified by silane [70, 71].

9.6   Cross-linking of Polyethylene by Microwave Energy

Cross-linking caused by light or other radiation is applicable only up to a certain layer size; the thickness of the part to be cross-linked is a process control factor for thermally induced free radical and silane cross-linking; expense and effort increase with increasing layer thickness. On the other hand, microwave cross-linking is independent of the thickness. This process is applicable to parts of any size.

The range between 10^9 and 10^{12} Hz of the electromagnetic spectrum is called the microwave range. In a high-frequency field the electrical behavior of polymers is determined mostly by the orientation of the molecular dipole in the electric field. In the IR region the dielectric state is determined by the absorption due to resonant vibrations of the lattice, by the valence, and by deformation oscillations [72]. Therefore, excitation of a polymer by microwaves differs from excitation by heating in the IR region. Only components with polar groups are excitable in a microwave field.

A polymer acts as a dielectric if placed in an electric field (for example, a capacitor) [72–74]. Dielectric losses enable the polymer to gain energy from the alternating field. The energy gain is given by the following equation:

$$N = E^2 \cdot 2 \cdot \pi \cdot f \cdot \varepsilon_r \cdot \tan \zeta$$

n   = loss or gain of energy
e   = field intensity
f   = frequency (Hz) of the alternating field
E_r = dielectric coefficient
$\tan \zeta$ = dielectric loss factor

The product $\varepsilon_r \cdot \tan \zeta$ is polymer-specific and determines the thermal behavior of the polymer. Table 48 lists ε_r and $\tan \zeta$ values of several polymers. Polymers are classified into two groups: those that are suitable as electrical insulators and those that exhibit dielectric loss in a high-frequency field. Since the values of $\tan \zeta$ and ε_r are finite for every polymer, it is theoretically possible to heat any polymer in a UHF field. From measurements of the thermal effect after a certain period of time in a UHF field it is seen that polymers which are suitable as electrical insulators show practically no heat gain at all. The border between polymers which are excitable in a UHF field and those that are not lies near values of $\tan \zeta = 1 \times 10^{-3}$, values differ only slightly for individual polymers.

Table 48   Dielectric coefficient and dielectric loss factor $\tan \zeta$ at 10^9 Hz and 23° for several compounds produced by BASF AG [75]

Polymer	ε_r	$\tan \zeta$
Polyethylene, Type Lupolen 1812 DSU	2.4	$5 \cdot 10^{-4}$
Polypropylene, Type Novolen 1320 HX	2.4	$1.5 \cdot 10^{-4}$
Polystyrene, Type 168 N	2.5	$2.5 \cdot 10^{-4}$
Polyoxymethylene, Type Ultraform N 2210	3.2	$8 \cdot 10^{-2}$
Polyamide, Type Ultramid A 3 K	3	$2 \cdot 10^{-2}$
Styrene acrylonitrite Copolymerisat, Type Luran 378 PG 7	3.1	$7.5 \cdot 10^{-3}$
Polyvinyl chloride, Type Vinoflex S 6115	2.8	$8 \cdot 10^{-3}$

Polymers with values of $\tan \zeta > 1 \times 10^{-3}$ can be heated in a UHF field. The welding (PVC) and vulcanization of polymers make use of that characteristic [76–78]. It seems to be impossible to use microwave energy for the processing of polyolefins, since their values for $\tan \zeta$ are extremely low. However, in practice these polymers are seldom used in the pure form; they are usually mixed with additives which are intensely polar and therefore allow the molding compound to be heated in a UHF field.

The same method allows modification of PE, so that cross-linking by a UHF field is possible. Since carbon black is a very desirable and suitable additive for many applications and facilitates the activation of PE in a UHF field it is used as a permanent UHF reactive additive [79, 80].

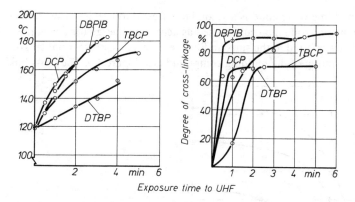

Figure 37 Cross-linking by microwave of PE with 5% (by weight) carbon black (Corax P) as UHF-active additive and 0.1 mole peroxide/1000 g PE, DCP = dicumyl peroxide, DBPIP = 1,3-di(tertiary-butyl peroxy isopropyl) benzene, TBCP = tertiary-butyl cumyl peroxide, DTBP = di-tertiary-butyl peroxide (2450 MHz, 2.1 kW) [79].

Several other methods use metallic powders as additives for PE [81]. The percentage of cross-linking is determined by the characteristic values of the UHF field strength, the amount and type of carbon black, and the type of peroxide (see Figure 37).

Heating a suitable polymer mixture by means of a UHF field has the distinct advantage of heating the molded article uniformly. While heating by heat conductivity results in the surface area being the hottest region, electrical heating will make it the coldest region since heat is lost to the surroundings. Therefore, very high temperatures of the heat transfer medium do not cause overheating of the surface area of the part. Cross-linking of the part takes place uniformly.

Cross-linking by UHF energy is not restricted to the use of carbon black as additive. The discussions of various methods of cross-linking PE have always included mention of the possibility that the process could be modified by simultaneous reaction with monomers. Triallyl oxy-s-triazine is a frequently used trifunctional monomer. This monomer is polar, just like all the other monomers which are described in the literature as suitable for cross-linking, and a mixture of PE and the monomer will become sufficiently excitable in a UHF field. The mixture can be heated and cross-linked by an UHF field [79].

The manufacture of electrical insulators is one of the main applications of cross-linked PE. The use of permanently UHF active additives may result in insulators with poor electrical properties; therefore, these additives are not used.

According to the literature [82], certain peroxides are themselves capable of absorbing energy from a UHF field [82]. During this process, energy is absorbed only at those points where it is needed to generate free radicals; therefore, in principle, heating of the polymer becomes superfluous. Thus, the peroxide acts as a UHF-active additive. It decomposes to free radicals by absorbing energy and is simultaneously converted to new chemical products. Especially suitable peroxides absorb energy very quickly from the UHF field and decompose to totally inactive or only slightly UHF-active fragments, while simultaneously cross-linking the polymer. UHF reactivity of the peroxide is therefore only temporary. The final products contain inactive decomposition products and exhibit excellent electrical properties. Tertiary butyl perbenzoate is an example of a suitable peroxide [82].

Heating of the molding compound is a secondary effect during microwave radiation. Energy absorbed by the peroxide molecule is used as activation energy for decomposition and is simultaneously transferred to the surrounding medium. Also, exothermal cross-linking contributes to the heating of the molding compound (see Figure 38). The cross-linking

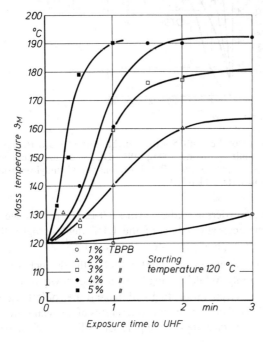

Figure 38
Bulk temperature as a function of exposure time and percentage of peroxide [83]. (TBPB = tertiary butyl perbenzoate.)

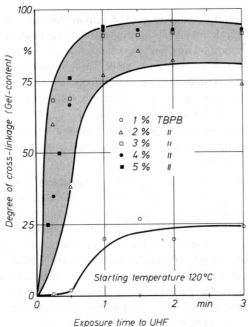

Figure 39
Degree of cross-linkage as a function of UHF exposure time and percentage of peroxide [83]. (TBPB = tertiary-butyl perbenzoate.)

reaction proceeds very quickly (see figure 39); the reaction requires as minimum temperature of 100 C because below this temperature the peroxide decomposes without producing any cross-linking (see Figure 40).

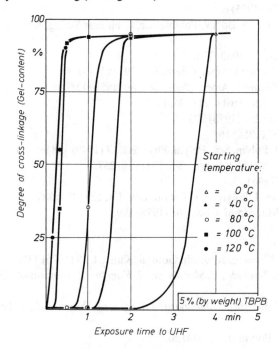

Figure 40
Degree of cross-linkage as a function of UHF exposure time [83].
[LDPE + 5% (by weight) tertiary-butyl perbenzoate]

Microwave cross-linking of PE was used commercially for the first time in the MDC process by Troester to manufacture insulators for high-voltage cables [84]. Bis(tertiary-butylperoxy)terephthalate serves as a suitable peroxide for this process [85].

Microwave cross-linking of PE takes place at relatively low bulk temperatures. The polyethylene strand is brought, without cooling, directly from the extruder into the UHF field; a moderate, but uniform, rise in temperature will occur. Thus, degradation of polymer due to thermal decomposition takes place to a much lesser degree than during other cross-linking methods.

An additional advantage is that water is not necessary as an energy transmitter and the process is carried out in a cross-linking tube filled with an inert gas.

Bibliography to Chapter 9

[1] *Saechtling, H.:* Kunststoff-Taschenbuch, 20. Ausgabe, Carl Hanser Verlag, München-Wien, 1977.
[2] DE-PS 1.164.645, W. R. Grace + Co.
[3] *Charlesby, A., M. G. Omerod:* Discourse at the International Symposium on free Radicals, Uppsala, 1961.
[4] *Bovery, F.A.:* »The Effects of Ionising Radiation on Natural and Synthetic High Polymers«, Interscience Publ., New York, 1958.

[5] *Chapiro, A.:* »Radiation Chemistry of Polymeric Systems«, Interscience Publ., London, 1962.

[6] *Charlesby, A.:* »Atomic Radiation and Polymers«, Pergamon Press, London, 1960.

[7] *Schandy, R.:* Österr. Chem. Z. 76 (1975) 6.

[8] *Wilski, H.:* in: Kunststoff-Handbuch, Bd. IV, Polyolefine, C. Hanser Verlag, München, 1969, 169.

[9] *Antonetty, G.:* BBC-Nachrichten 51 (1969) 274.

[10] *Dole, M., D. C. Milner, T. F. Williams:* J. Amer. Chem. Soc. 79 (1957) 4809.

[11] *Dole, M., D. C. Milner, T. F. Williams:* J. Amer. Chem. Soc. 80 (1958) 1580.

[12] *Lyons, B. J., A. S. Fox:* J. Polym. Sci., Part C 21 (1967) 159.

[13] *Wilski, H.:* Kolloid-Z-. Z. Polym. 251 (1973) 703.

[14] *Schumacher, K.:* Kolloid-Z., 157 (1958) 16.

[15] *Yamamoto, F., S. Yamakawa:* J. Polym. Sci., Polym. Phys. Ed. 17 (1979) 1581.

[16] *Lawton, E. J., J. S. Balwit, R. S. Powell:* J. Polym. Sci. 32 (1958) 257.

[17] *Zyball, A.:* Kunststoffe 67 (1977) 461.

[18] *Singer, K., M. Joshi, J. Silverman:* J. Polym. Sci., Polym. Lett. Ed. 12 (1974) 387.

[19] *Lee, D. W., D. Braun:* Angew. Makromol. Chem. 68 (1978) 199.

[20] GB-PS 1.095.772, Raychem Ltd.

[21] DE-OS 2.420.784, Siemens AG.

[22] *Ahne, H., R. Wiedenmann, W. Kleeberg:* Kautsch. Gummi, Kunstst. 28 (1975) 135.

[23] *Hagiwara, M., T. Tagawa, E. Tsuchida, I. Shinohara, T. Kagiya:* J. Macromol. Sci. Chem. 7 (1973) 1591.

[24] *Hagiwara, M., T. Tagawa, E. Tsuchida, I. Shinobara, T. Kagiya:* J. Polym. Sci., Polym. Lett. Ed. 11 (1973) 613.

[25] *Mitsui, H., F. Hosoi, T. Kagiya:* Polym. J. 6 (1974) 20.

[26] *Mitsui, H., F. Hosoi, M. Ushirokawa:* J. Appl. Polym. Sci. 19 (1975) 361.

[27] *Köhnlein, E.:* Kunststoffe 65 (1975) 583.

[28] *Patel, G. N., L. D. Jlario, A. Keller, E. Martuscelli:* Makromol. Chem. 175 (1974) 983.

[29] *Wilski, H.:* Angew. Chemie 71 (1959) 612.

[30] US-PS 3.219.566, Dow Chemical Co.

[31] US-PS 2.484.529, DuPont.

[32] *Oster, G., G. K. Oster, H. Moroson:* J. Polym. Sci. 34 (1959) 671.

[33] DE-OS 2.337.813, Ciba Geigy AG.

[34] *Soumelis, K., H. Wilski:* Kunststoffe 52 (1962) 471.

[35] GB-PS 1.131.010, Raychem Corp.

[36] GB-PS 1.048.067, W. R. Grace & Co.

[37] GB-PS 1.087.403, Raychem Corp.

[38] Unpublished report at the Institut für Kunststoffverarbeitung, TH Aachen.

[39] *Braun, D., W. Brendlein:* Kunststofftechnik 9 (1970) 275.

[40] *Rauer, K., J. Groepper:* Kunststoffe 62 (1972) 699.

[41] US-P 3.242.159, Pullman Inc.

[42] *van Drumpt, J. D., H. H. J. Oosterwijk:* J. Polym. Sci., Polym. Chem. Ed. 14 (1976) 1495.

[43] D-OS 2.245.552, Ruhrchemie AG.

[44] *Dannenberg, E. M., M. E. Jordan, H. M. Cole:* J. Polym. Sci. 31 (1958) 127.

[45] *Stemmer, H. D.:* Kautsch. Gummi, Kunstst. 14 (1961) 146.

[46] *Markert, H., R. Wiedenmann:* Discourse at the Hauptversammlung der Gesellschaft Deutscher Chemiker in Karlsruhe, Sept. 1971.

[47] *Rheinfeld, D.:* Dissertation at the Institut für Kunststoffverarbeitung, RWTH Aachen, 1972.

[48] *Rheinfeld, D.:* Gummi, Asbest, Kunstst. a) 28 (1975) 80; b) 28 (1975) 390; c) 28 (1975) 452.

[49] *Wunsch, K., G. Kohlmann:* Plaste Kautsch. 13 (1966) 258.

[50] DE-OS 2.308.576, Siemens AG.

[51] DE-OS 1.494.107, Farbwerke Hoechst AG.

[52] *Barton, J., J. Pavlinec:* Plaste Kautsch. 15 (1968) 397.

[53] *Otani, K.:* Jpn. Plast. Age, a) 16,2 (1978) 21; b) 16,3 (1978) 33; c) 16,4 (1978) 19.

[54] FR-PS 2.207.944, General Electric Co.

[55] *Franzkoch, B.:* Dissertation at the Institut für Kunststoffverarbeitung, RWTH Aachen, 1979.

[56] DE-OS 1.915.033, Vereinigte Deutsche Metallwerke AG.

[57] *Engel, Th.:* Kunststoffe 57 (1967) 536.

[58] US-PS 3.484.352, Phillips Petroleum Co.

[59] DE-AS 1.694.130, The Furukawa Electric Co.

[60] DE-AS 2.348.468, The Furukawa Electric Co.

[61] DE-OS 2.350.876, Dow Corning Corp.

[62] DE-OS 2.255.116, Dow Corning Corp.

[63] DE-OS 2.406.844, Dow Corning Corp.

[64] *Scott, H., J. F. Humphries:* Mod. Plast. 50 (1973) 82.

[65] DE-OS 26 49 874, Kabel- u. Metallwerke Gutehoffnungshütte AG.

[66] DE-OS 26 46 080, Kabel- u. Metallwerke Gutehoffnungshütte AG.

[67] JP-PS 76-61.549, Furukawa Electric Co., Ltd.

[68] FR-PS 1.577.875, Midland Silicone Ltd.

[69] *Voigt, H. U.:* Kautsch. Gummi, Kunstst. 29 (1976) 17.

[70] DE-OS 26 11 491, Sekisui Kagaku Kogyo, K. K., Japan.

[71] DE-OS 2.451.218, Bayer AG.

[72] *Amrhein, E.:* Kolloid-Z., Z. Polym. 216–217 (1967) 38.

[73] *Püschner, H.:* »Wärme durch Mikrowellen«, Philips Techn. Bibl., Philips Eindhoven, 1964.

[74] *Bergmann, K.:* Kunststoffe 61 (1971) 226.

[75] Company report, BASF AG »Kunststoffe für die Elektrotechnik«.

[76] *Schindler, U.:* Gummi, Asbest, Kunstst. 31 (1978) 222.

[77] *Focht, H.:* Kautsch., Gummi, Kunstst. 29 (1976) 272.

[78] *Focht, H.:* Gummi, Asbest, Kunstst. 32 (1979) 622.

[79] *Menges, G., K. Kircher, B. Franzkoch, W. Hoffacker:* Kunststoffe 69 (1979) 430.

[80] US-PS 3.755.519, Beaunit Corp., New York.

[81] US-PS 2.972.780, Cabot Corp.

[82] a) DE-OS 26 11 349, b) DE-OS 28 03 252, c) DE-OS 29 04 086, Vereinigung zur Förderung des Instituts für Kunststoffverarbeitung in Industrie und Handwerk an der RWTH Aachen.

[83] *Menges, G., K. Kircher, B. Franzkoch:* Kunststoffe 70 (1980) 45.

[84] *N. N.:* Company report Paul Troester, Hannover.

[85] Information for the development product D 165, Akzochemie.

10　Cross-linking of other Polymers

10.1　Polypropylene

Cross-linking of polypropylene (PP) is of lesser commercial importance than the cross-linking of lower density polyethylene (LDPE). Chain degradation does not occur very frequently during cross-linking of LDPE; but cross-linking of PP is generally accompanied by a high degree of chain degradation.

The high percentage of tertiary hydrogen atoms in polypropylene is the main reason for its evidently easy chain degradation during free-radical-initiated cross-linking. First, primary radicals initiate abstraction of a tertiary hydrogen atom which will result in formation of PP radicals:

$$-CH_2-CH-CH_2-\overset{\overset{H}{|}}{C}-CH_2- \quad + \quad \bullet O-R \quad \longrightarrow \quad -CH_2-CH-CH_2-\overset{\bullet}{C}-CH_2- \quad + \quad ROH$$
$$\underset{CH_3}{|} \qquad \underset{CH_3}{|} \qquad\qquad\qquad\qquad\qquad \underset{CH_3}{|} \qquad \underset{CH_3}{|}$$

     Polypropylene　　　　　　　　　　　Primary　　　　　　Polypropylene radical

　　　　　　　　　　　　　　　　　　　radical

Tertiary hydrogen atoms have a lower bond energy than primary hydrogen atoms; therefore, chain radicals are formed much faster in PP than in PE. The lower bond energy corresponds to a greater stability of the chain radicals, and dimerization is less likely to happen than in linear PE. The higher stability of PP chain radicals allows side reactions such as inactivation by reacting with primary radicals and by chain cleavage.

Reaction of PP radical with primary radical:

$$-CH_2-CH-CH_2-\overset{\bullet}{C}-CH_2-CH- \quad + \quad R-O\bullet \quad \longrightarrow \quad -CH_2-CH-CH_2-\overset{\overset{R}{\overset{|}{O}}}{\underset{|}{C}}-CH_2-CH-$$
$$\underset{CH_3}{|} \qquad \underset{CH_3}{|} \qquad \underset{CH_3}{|} \qquad\qquad\qquad\qquad \underset{CH_3}{|} \qquad \underset{CH_3}{|} \qquad \underset{CH_3}{|}$$

Splitting of the PP radical:

$$-CH_2-CH-CH_2-\overset{\bullet}{C}-CH_2-CH- \quad \longrightarrow \quad -CH_2-CH-CH=CH \quad + \quad \bullet CH_2-CH-$$
$$\underset{CH_3}{|} \qquad \underset{CH_3}{|} \qquad \underset{CH_3}{|} \qquad\qquad\qquad \underset{CH_3}{|} \qquad \underset{CH_3}{|} \qquad \underset{CH_3}{|}$$

$$\bullet CH_2-CH- \quad \longrightarrow \quad CH_3-\overset{\bullet}{C}-$$
$$\underset{CH_3}{|} \qquad\qquad\qquad \underset{CH_3}{|}$$

On the one hand, the cross-linking process will yield a modified polymer by formation of new bonds, and on the other hand it will lead to undesirable deterioration by chain depolymerization. Only a few literatures references deal exclusively with cross-linking of PP [1, 2].

10.2   Ethylene Vinyl Acetate Copolymers

Compared to the cross-linking of PP, cross-linking of this copolymer is accompanied by very little degradation. The effectiveness of the free radicals is higher for the cross-linking of ethylene vinyl acetate copolymer than for the cross-linking of LDPE [3]. The use of polyfunctionally unsaturated monomers, such as triallyl cyanurate, during cross-linking of ethylene vinyl acetate copolymers will result in a higher degree of cross-linking [4, 5].

10.3   Polyisobutylene

Pure polyisobutylene cannot be cross-linked by free radicals; the chain depolymerization reaction occurs more frequently than the cross-linking reaction, thus resulting in poor product quality. According to [6], primary free radicals abstract the less active hydrogen atoms from a methyl group (Eq. a). Immediately, the newly formed polymer radicals disproportionate by chain cleavage and form an olefin and a new polymer radical containing a radical site in the end group (Eq. b); the radical reacts further by additional chain depolymerization (Eq. c):

$$
-CH_2-\underset{\underset{CH_3}{|}}{\overset{\overset{CH_3}{|}}{C}}-CH_2-\underset{\underset{CH_3}{|}}{\overset{\overset{CH_3}{|}}{C}}-CH_2- \;+\; R-O\cdot \;\longrightarrow\; -CH_2-\underset{\underset{\overset{CH_2}{\cdot}}{|}}{\overset{\overset{CH_3}{|}}{C}}-CH_2-\underset{\underset{CH_3}{|}}{\overset{\overset{CH_3}{|}}{C}}-CH_2- \;+\; R-OH \qquad (a)
$$

$$
-CH_2-\underset{\underset{\overset{CH_2}{\cdot}}{|}}{\overset{\overset{CH_3}{|}}{C}}-CH_2-\underset{\underset{CH_3}{|}}{\overset{\overset{CH_3}{|}}{C}}-CH_2- \;\longrightarrow\; -CH_2-\underset{\underset{CH_2}{\|}}{\overset{\overset{CH_3}{|}}{C}} \;+\; \cdot CH_2-\underset{\underset{CH_3}{|}}{\overset{\overset{CH_3}{|}}{C}}-CH_2- \qquad (b)
$$

$$
\cdot CH_2-\underset{\underset{CH_3}{|}}{\overset{\overset{CH_3}{|}}{C}}-CH_2- \;\longrightarrow\; \cdot CH_3 \;+\; CH_2=\underset{\underset{CH_3}{|}}{\overset{\overset{CH_3}{|}}{C}}-CH_2- \qquad (c)
$$

The cross-linking of modified polyisobutylene to butyl rubber is an important industrial process. This is not a cross-linking reaction induced by free radicals. The reaction utilizes the small amount of available double bonds for chemical cross-linking with sulfur. Cross-linking with sulfur requires an accelerator, such as the accelerator mix containing 60% tetramethyl thiuram disulfide and 40% mercaptobenzothiazole combined with ZnO.

Another method uses p-benzoquinone dioxime to cross-link butyl rubber. This reaction proceeds very rapidly if sulfur and sulfur accelerators are present. The mechanism of cross-linking with sulfur is described in detail in chapter 11.

10.4   Fluoro Polymers

Perfluorinated polymers cannot be cross-linked by free radicals, since they depolymerize at the temperatures used in that reaction. However, fluoro polymers containing hydrogen atoms can be cross-linked with either peroxides or diamines. These fluoro polymers are important to the manufacture of fluoro elastomers. Copolymers of vinylidene fluoride and hexafluoropropylene are the most important representative of that class.

$$- (- CF_2 - CH_2 -)_n - (- \overset{\overset{\displaystyle CF_3}{|}}{CF} - CF_2)_m -$$

Vinylidene fluoridehexafluoropropylene copolymers

During peroxide-induced cross-linking, fluoro polymers react with primary free radicals, and the newly formed polymer radicals dimerize (similar to the cross-linking of PE) [7]:

$$- CF_2 - CH_2 - \overset{\overset{\displaystyle CF_3}{|}}{CF} - CF_2 - \ + \ R - O\bullet \ \longrightarrow \ - CF_2 - \overset{\bullet}{C}H - \overset{\overset{\displaystyle CF_3}{|}}{CF} - CF_2 - \ + \ R - OH$$

$$- CF_2 - \overset{\bullet}{C}H - \overset{\overset{\displaystyle CF_3}{|}}{CF} - CF_2 -$$

$$+$$

$$- CF_2 - \overset{\bullet}{C}H - \overset{\overset{\displaystyle CF_3}{|}}{CF} - CF_2$$

$$\longrightarrow$$

$$\begin{array}{c} - CF_2 - CH - \overset{\overset{\displaystyle CF_3}{|}}{CF} - CF_2 - \\ | \\ - CF_2 - CH - \overset{\underset{\displaystyle CF_3}{|}}{CF} - CF_2 - \end{array}$$

A secondary reaction eliminates hydrogen fluoride, with subsequent formation of carbon chain double bonds:

$$- CF_2 - CH_2 - \overset{\overset{\displaystyle CF_3}{|}}{CF} - CF_2 - \ \longrightarrow \ HF \ + \ - CF_2 - CH = \overset{\overset{\displaystyle CF_3}{|}}{C} - CF_2 -$$

It becomes necessary to add a base in order to neutralize the very corrosive, acidic hydrogen fluoride; thus, MgO is added to the cross-linking process:

$$MgO \ + \ 2 \ HF \ \longrightarrow \ MgF_2 \ + \ H_2O$$

The relatively easy formation of double bonds allows cross-linking with diamines:

$$\begin{array}{c} - CF_2 - CH = \overset{\overset{\displaystyle CF_3}{|}}{C} - CF_2 - \\ \\ NH_2 \\ | \\ + \quad R \\ | \\ NH_2 \\ \\ - CF_2 - CH = \overset{\underset{\displaystyle CF_3}{|}}{C} - CF_2 - \end{array} \quad \longrightarrow \quad \begin{array}{c} - CF_2 - CH_2 - \overset{\overset{\displaystyle CF_3}{|}}{C} - CF_2 - \\ | \\ NH \\ | \\ R \\ | \\ NH \\ | \\ - CF_2 - CH_2 - \overset{\underset{\displaystyle CF_3}{|}}{C} - CF_2 - \end{array}$$

The excess amino compound is a suitable neutralizer for the generated hydrogen fluoride.

$$HF \ + \ NH_2-R-NH_2 \ \longrightarrow \ \left[\overset{+}{N}H_3-R-NH_2 \right] \ F^-$$

It is conceivable that the elimination of hydrogen fluoride from the fluoro polymer is an equilibrium reaction which is influenced by the basic character of the diamine.

10.5 Thermoplastic Polyurethanes

Reaction casting of intermediates such as polyol and di- or polyisocyanate is the method most widely used to manufacture cross-linked polyurethane (PUR) (see section 8.1). Furthermore, it may be possible to process thermoplastic linear PUR in screw extruders with subsequent cross-linking of the extruded parts. Thus, synthesis of PUR has to be performed by the raw material manufacture while cross-linking of thermoplastic polymers is carried out in the processing plant.

Several methods can be used to cross-link thermoplastic PUR. One process utilizes the reaction of the hydroxyl groups, present in thermoplastic PUR with diisocyanates. This reaction corresponds to the synthesis of polyurethane.

The reaction between uretdione groups, which are part of the chain, and hydroxyl groups represents an additional cross-linking method.

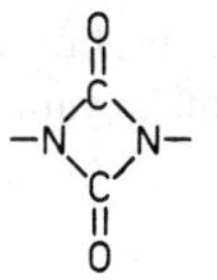

Uretdione group

A temperature of 120 °C is necessary to activate the uretdion group (from dimerized diisocyanate) to the point where reaction with hydroxyl groups becomes feasible [8–10]:

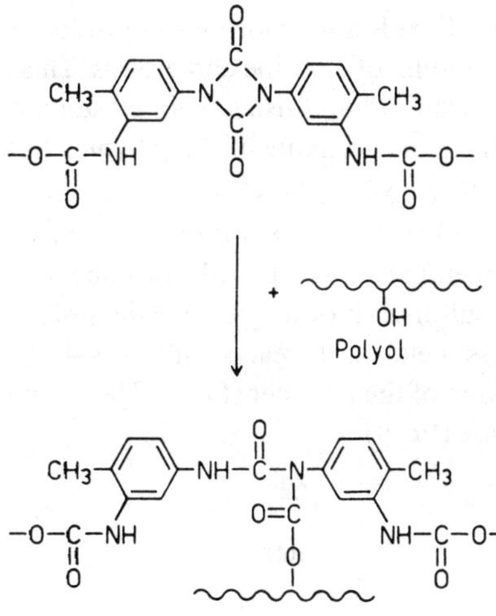

PUR made from the basic intermediate p,p′-diphenylmethane diisocyanate (MDI) [11, 12] can be cross-linked by free radicals. Reaction between primary free radicals and the methylene group of the diphenylmethane group results in elimination of hydrogen and generates chain radicals which dimerize to tetraphenylethane compounds:

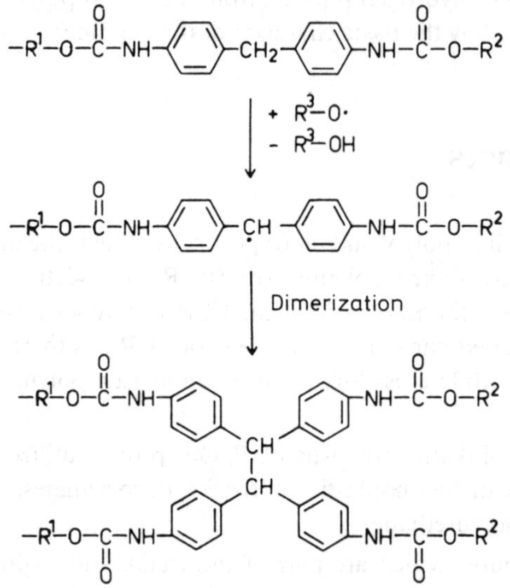

Peroxide-induced cross-linking is applicable to polyester urethanes and to polyether urethanes; however, polyethers cannot be branched. Peroxides will cleave the branched polyethers [13]. The use of polyfunctional unsaturated monomers, such as triallyl cyanurate, makes acceleration of the cross-linking process possible [14].

10.6   Polyethylene Sulfonyl Chloride

Treating polyethylene with a mixture of sulfur dioxide and chlorine will produce polyethylene sulfonyl chloride which contains a small amount of chlorine end groups. This reaction involves free radicals. The modified PE is vulcanized and used as an elastomer. The sulfonyl chloride groups are involved in the chemical cross-linking of the polymer. Synthesis of polyethylene sulfonyl chloride is the job of the polymer manufacture.

Polyethylene sulfonyl chloride can be cross-linked by a mixture of metal oxide and carboxylic acid [6]. In a preliminary reaction the metal oxide reacts with carbonic acid and forms water (Eq. a); water will then hydrolyze the sulfonyl chloride group to the sulfonic acid group and hydrogen chloride (Eq. b). The excess metal oxide reacts with the sulfonic acid, resulting in elimination of water and cross-linking of the polymer (Eq. c). The eliminated hydrogen chloride is neutralized by the metal oxide (Eq. d).

$$ZnO \ + \ 2\ C_{17}H_{35} - COOH \ \longrightarrow \ (C_{17}H_{35} - COO^-)_2\ Zn^{2+} \ + \ H_2O \qquad (a)$$

| Metal oxide | Stearic acid | Salt | Water |

$$H_2O \; + \; -CH_2-CH_2-\underset{\underset{Cl}{|}}{\underset{SO_2}{|}}{\overset{|}{CH}}-CH_2-CH_2- \;\longrightarrow\; HCl \; + \; -CH_2-CH_2-\underset{\underset{OH}{|}}{\underset{SO_2}{|}}{\overset{|}{CH}}-(CH_2)_2- \qquad (b)$$

Sulfonyl chloride group   Sulfonic acid group

$$2 \; -CH_2-CH_2-\underset{\underset{OH}{|}}{\underset{SO_2}{|}}{\overset{|}{CH}}-CH_2-CH_2- \; + \; ZnO \;\longrightarrow\; \begin{array}{c} -CH_2-CH_2-CH-CH_2-CH_2- \\ | \\ SO_3^- \\ | \\ Zn^{2+} \\ | \\ SO_3^- \\ | \\ -CH_2-CH_2-CH-CH_2-CH_2- \end{array} \; + \; H_2O \qquad (c)$$

$$2 \; HCl \; + \; ZnO \;\longrightarrow\; ZnCl_2 \; + \; H_2O \qquad (d)$$

A secondary reaction occurs during the heat treatment of polyethylene sulfonyl chloride, resulting in the elimination of SO_2 and HCl.

Polyethylene sulfonyl chloride can also be cross-linked by diepoxides [6]:

$$\begin{array}{c} | \\ CH_2 \\ | \\ CH-SO_2-Cl \\ | \\ CH_2 \\ | \end{array} \; + \; \underset{\text{Diepoxide}}{CH_2-CH-R-CH-CH_2} \; + \; \begin{array}{c} | \\ CH_2 \\ | \\ Cl-SO_2-CH \\ | \\ CH_2 \\ | \end{array} \;\longrightarrow$$

Sulfonyl chloride

$$\begin{array}{c} | \\ CH_2 \\ | \\ CH-SO_2-O-CH-R-CH-O-SO_2-CH \\ | \qquad\qquad | \qquad\quad | \qquad\qquad | \\ CH_2 \qquad\quad CH_2 \quad CH_2 \qquad\quad CH_2 \\ | \qquad\qquad | \qquad\quad | \qquad\qquad | \\ \qquad\qquad Cl \qquad\quad Cl \end{array}$$

Ester of sulfonic acid
(Cross-linked polymer)

Another cross-linking method uses diols, such as ethylene glycol, hydroquinone, or pentaerythritol [6]:

$$\begin{array}{c} | \\ CH_2 \\ | \\ CH-SO_2-Cl \\ | \\ CH_2 \\ | \end{array} \; + \; \underset{\text{Ethylene glycol}}{HO-CH_2-CH_2-OH} \; + \; \begin{array}{c} | \\ CH_2 \\ | \\ Cl-SO_2-CH \\ | \\ CH_2 \\ | \end{array} \;\longrightarrow$$

$$\begin{array}{c} | \\ CH_2 \\ | \\ CH-SO_2-O-CH_2-CH_2-O-SO_2-CH \\ | \qquad\qquad\qquad\qquad\qquad\qquad | \\ CH_2 \qquad\qquad\qquad\qquad\qquad CH_2 \\ | \qquad\qquad\qquad\qquad\qquad\qquad | \end{array} \; + \; 2 \; HCl$$

Ester of sulfonic acid
(Cross-linked polymer)

Hydrogen chloride, eliminated during cross-linking, is neutralized by MgO producing magnesium chloride and water.

The cross-linking by diamines occurs according to a very similar reaction mechanism:

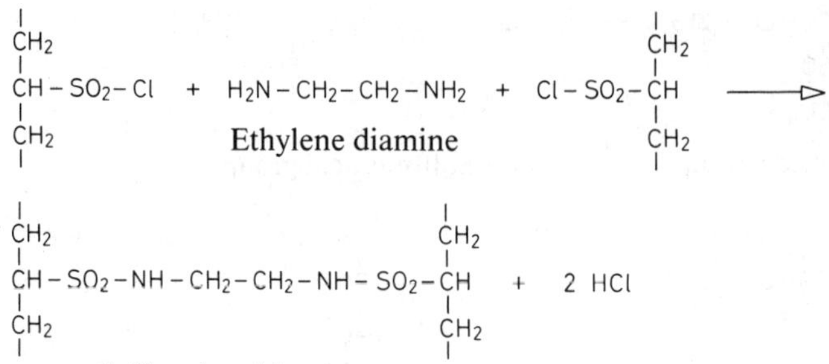

$$
\begin{array}{ccc}
\overset{|}{CH_2} & & \overset{|}{CH_2} \\
\overset{|}{CH} - SO_2 - Cl \quad + \quad H_2N - CH_2 - CH_2 - NH_2 \quad + \quad Cl - SO_2 - \overset{|}{CH} & \longrightarrow \\
\overset{|}{CH_2} & \text{Ethylene diamine} & \overset{|}{CH_2} \\
| & & |
\end{array}
$$

$$
\begin{array}{cc}
\overset{|}{CH_2} & \overset{|}{CH_2} \\
\overset{|}{CH} - SO_2 - NH - CH_2 - CH_2 - NH - SO_2 - \overset{|}{CH} \quad + \quad 2\ HCl \\
\overset{|}{CH_2} & \overset{|}{CH_2} \\
| & |
\end{array}
$$

Sulfonyl amide
(Cross-linked polymer)

The eliminated hydrogen chloride is neutralized by an excess of amine and forms ammonium salts. The reaction between the sulfonyl chloride group and the amine is so fast that this method is only of limited use to industrial applications. Therefore, less active compounds have to be used, such as substituted quinoxalines, which react only in a tautomeric form [15]:

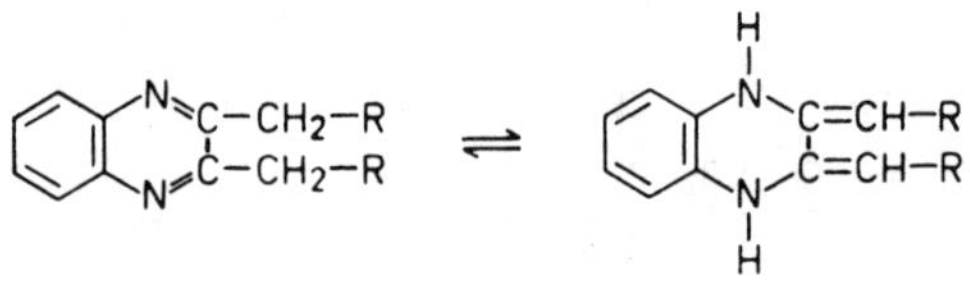

Substituted quinoxaline   Tautomeric form

10.7   Polychloroprene

Polychloroprene is a polyene (such as natural rubber and polybutadiene) and as such can be cross-linked by sulfur; but this method is of little importance.

$$
\left(CH_2 - C = CH - CH_2 \right)_n \quad
$$

Poly-2-chlorobutadiene
(Polychloroprene)

(with Cl on the C)

The most frequently used cross-linking methods of poylchloroprene use either metal oxides or 2-mercaptoimidazole as cross-linking agents.

Polymerization of 2-chlorobutadiene ($CH_2 = \overset{Cl}{\underset{|}{C}} - CH = CH_2$) will produce primarily a linear polymer with chains containing double bonds as a result of 1,4 linkage (see above). Approximately (1,5% (mole/mole) monomer reacts in 1,2 linkage; thus resulting in branches of vinyl groups instead of (ethylenic) unsaturated chain structures.

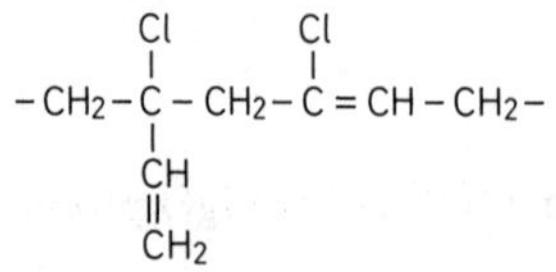

$$
\begin{array}{cc}
\overset{Cl}{|} & \overset{Cl}{|} \\
- CH_2 - \overset{|}{C} - CH_2 - \overset{|}{C} = CH - CH_2 - \\
\overset{|}{CH} & \\
\overset{\parallel}{CH_2} &
\end{array}
$$

1,2 Linkage   1,4 Linkage

The most active group in polychloroprene is the vinyl side chain containing a chlorine atom in the alpha position. The vinyl group containing the chlorine atom in the alpha position is thought to be able to convert to a tautomeric form which acts as the key component for the cross-linking reaction:

$$
\begin{array}{ccc}
& \text{Cl} & \\
& | & \\
-\text{CH}_2-\text{C}- & & -\text{CH}_2-\text{C}- \\
| & \rightleftharpoons & \| \\
\text{CH} & & \text{CH} \\
\| & & | \\
\text{CH}_2 & & \text{CH}_2 \\
& & | \\
& & \text{Cl}
\end{array}
$$

Treatment of polychloroprene with an excess of amine, for instance piperidine, results in the following reaction. It is believed that the tautomeric form is responsible for the reaction mechanism [16]:

$$
\begin{array}{c}
\text{CH}_2 \\
| \\
\text{Cl}-\text{C}-\text{CH}=\text{CH}_2 \\
| \\
\text{CH}_2 \\
|
\end{array}
\longrightarrow
\begin{array}{c}
\text{CH}_2 \\
| \\
\text{C}=\text{CH}-\text{CH}_2-\text{Cl} \\
| \\
\text{CH}_2 \\
|
\end{array}
+ \ \text{HN}
\begin{array}{c}
\diagup\text{CH}_2-\text{CH}_2\diagdown \\
\qquad\qquad\quad\text{CH}_2 \\
\diagdown\text{CH}_2-\text{CH}_2\diagup
\end{array}
\longrightarrow
$$

$$
\begin{array}{c}
\text{CH}_2 \\
| \\
\text{C}=\text{CH}-\text{CH}_2-\text{N}
\begin{array}{c}
\diagup\text{CH}_2-\text{CH}_2\diagdown \\
\qquad\qquad\quad\text{CH}_2 \\
\diagdown\text{CH}_2-\text{CH}_2\diagup
\end{array}
\ + \ \text{HCl} \\
| \\
\text{CH}_2 \\
|
\end{array}
$$

Vulcanization of polychloroprene with diamines proceeds according to a similar reaction mechanism; excess amine will neutralize the hydrogen chloride that is generated.

$$
\begin{array}{c}
\text{CH}_2 \\
| \\
\text{Cl}-\text{C}-\text{CH}=\text{CH}_2 \\
| \\
\text{CH}_2 \\
|
\end{array}
+ \ \text{HN}
\begin{array}{c}
\diagup\text{CH}_2-\text{CH}_2\diagdown \\
\qquad\qquad\quad\text{NH} \\
\diagdown\text{CH}_2-\text{CH}_2\diagup
\end{array}
+ \ \text{CH}_2=\text{CH}-
\begin{array}{c}
\text{CH}_2 \\
| \\
\text{C}-\text{Cl} \\
| \\
\text{CH}_2 \\
|
\end{array}
\longrightarrow
$$

$$
\begin{array}{c}
\text{CH}_2 \\
| \\
\text{C}=\text{CH}-\text{CH}_2-\text{N}
\begin{array}{c}
\diagup\text{CH}_2-\text{CH}_2\diagdown \\
\qquad\qquad\quad\text{N}-\text{CH}_2-\text{CH}=
\end{array}
\begin{array}{c}
\text{CH}_2 \\
| \\
\text{C} \\
| \\
\text{CH}_2 \\
|
\end{array}
\ + \ 2 \ \text{HCl} \\
| \\
\text{CH}_2 \\
|
\end{array}
$$

Likewise, vulcanization by metal oxides involves the chlorine atom in the reaction (see next page). A mixture of MgO and ZnO will give the best results with respect to ease of cross-linking and properties of the final product [6].

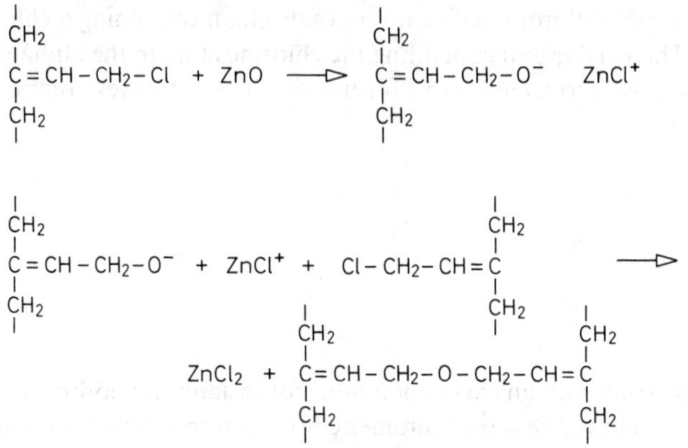

$$
\begin{array}{c}
| \\
CH_2 \\
| \\
C = CH - CH_2 - Cl \quad + \quad ZnO \quad \longrightarrow \quad C = CH - CH_2 - O^- \quad ZnCl^+ \\
| \\
CH_2 \\
|
\end{array}
\qquad
\begin{array}{c}
| \\
CH_2 \\
| \\
C = CH - CH_2 - O^- \\
| \\
CH_2 \\
|
\end{array}
$$

$$
\begin{array}{c}
| \\
CH_2 \\
| \\
C = CH - CH_2 - O^- \quad + \quad ZnCl^+ \quad + \quad Cl - CH_2 - CH = C \\
| \\
CH_2 \\
|
\end{array}
\qquad
\begin{array}{c}
| \\
CH_2 \\
| \\
\\
| \\
CH_2 \quad CH_2 \\
|
\end{array}
\longrightarrow
$$

$$
ZnCl_2 \quad + \quad
\begin{array}{c}
CH_2 \qquad\qquad CH_2 \\
| \qquad\qquad\qquad | \\
C = CH - CH_2 - O - CH_2 - CH = C \\
| \qquad\qquad\qquad | \\
CH_2 \qquad\qquad CH_2 \\
|
\end{array}
$$

During the cross-linking of polychloroprene by 2-mercaptoimidazol, the reactive chloro compound is initially added to the thiourea group of 2-mercaptoimidazol [17]:

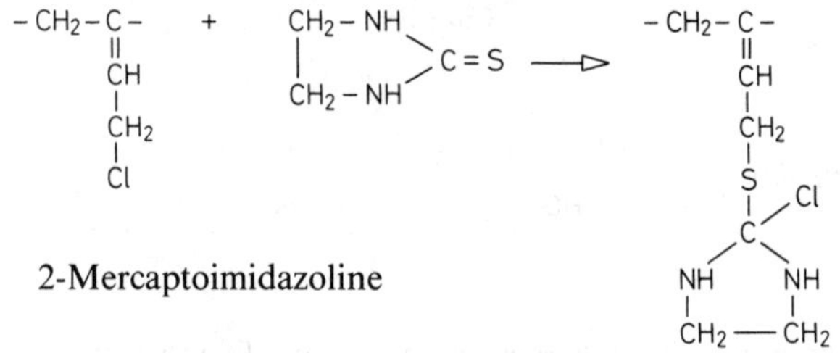

$$
\begin{array}{c}
- CH_2 - C - \\
\parallel \\
CH \\
| \\
CH_2 \\
| \\
Cl
\end{array}
\quad + \quad
\begin{array}{c}
CH_2 - NH \\
| \qquad\qquad\; C = S \\
CH_2 - NH
\end{array}
\longrightarrow
\begin{array}{c}
- CH_2 - C - \\
\parallel \\
CH \\
| \\
CH_2 \\
| \\
S \quad Cl \\
C \\
NH \quad NH \\
| \qquad | \\
CH_2 \!-\! CH_2
\end{array}
$$

2-Mercaptoimidazoline

ZnO causes further reaction of the intermediate compound, resulting in the cross-linking of polychloroprene. The following cross-linking site is formed:

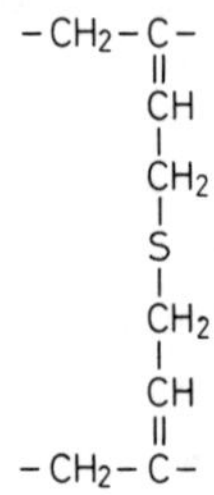

$$
\begin{array}{c}
- CH_2 - C - \\
\parallel \\
CH \\
| \\
CH_2 \\
| \\
S \\
| \\
CH_2 \\
| \\
CH \\
\parallel \\
- CH_2 - C -
\end{array}
$$

Cross-linked polychloroprene
(Cross-linking agent: 2-mercaptoimidazol + MgO)

10.8   Silicones

Silicones or organo polysiloxanes, respectively, are synthesized from organochlorosilanes. These are then hydrolyzed with water to silanols and are subsequently polymerized by polycondensation. The most important organochlorosilanes are monoalkyl trichlorosilane, dialkyl dichlorosilane, and trialkyl monochlorosilane:

$$R - \overset{\overset{\displaystyle Cl}{|}}{\underset{\underset{\displaystyle Cl}{|}}{Si}} - Cl \quad + \quad 3\ H_2O \quad \longrightarrow \quad R - \overset{\overset{\displaystyle OH}{|}}{\underset{\underset{\displaystyle OH}{|}}{Si}} - OH$$

Monoalkyl
trichlorosilane

Monoalkylsilanol

$$Cl - \overset{\overset{\displaystyle R}{|}}{\underset{\underset{\displaystyle R}{|}}{Si}} - Cl \quad + \quad 2\ H_2O \quad \longrightarrow \quad HO - \overset{\overset{\displaystyle R}{|}}{\underset{\underset{\displaystyle R}{|}}{Si}} - OH$$

Dialkyl
dichlorosilane

Dialkyl silanol

$$R - \overset{\overset{\displaystyle R}{|}}{\underset{\underset{\displaystyle R}{|}}{Si}} - Cl \quad + \quad H_2O \quad \longrightarrow \quad R - \overset{\overset{\displaystyle R}{|}}{\underset{\underset{\displaystyle R}{|}}{Si}} - OH$$

Trialkyl
monochlorosilane

Trialkyl silanol

Polycondensation of dialkyl silanes produces linear macromolecules:

$$n\ HO - \overset{\overset{\displaystyle R}{|}}{\underset{\underset{\displaystyle R}{|}}{Si}} - OH \quad \longrightarrow \quad HO - \overset{\overset{\displaystyle R}{|}}{\underset{\underset{\displaystyle R}{|}}{Si}} - O \left[- \overset{\overset{\displaystyle R}{|}}{\underset{\underset{\displaystyle R}{|}}{Si}} - O \right]_{n-2} - \overset{\overset{\displaystyle R}{|}}{\underset{\underset{\displaystyle R}{|}}{Si}} - OH \quad + \quad (n-1)\ H_2O$$

Trialkyl silanes will terminate the polycondensation and should be used only in small amounts. The size of the polycondensate molecule is determined by the ratio of di- to trialkyl silanol.

$$n\ HO - \overset{\overset{\displaystyle R}{|}}{\underset{\underset{\displaystyle R}{|}}{Si}} - OH \quad + \quad 2\ R - \overset{\overset{\displaystyle R}{|}}{\underset{\underset{\displaystyle R}{|}}{Si}} - OH \quad \longrightarrow \quad R - \overset{\overset{\displaystyle R}{|}}{\underset{\underset{\displaystyle R}{|}}{Si}} + O - \overset{\overset{\displaystyle R}{|}}{\underset{\underset{\displaystyle R}{|}}{Si}} \underset{n}{\Big]} O - \overset{\overset{\displaystyle R}{|}}{\underset{\underset{\displaystyle R}{|}}{Si}} - R$$

Linear silicone

Monoalkyl silane can function as the cross-linking agent. Polycondensations using large amounts of monoalkyl silane produce a cross-linked polymer which cannot be processed further; it has the following chemical structure:

$$-O - \overset{\overset{\displaystyle R}{|}}{\underset{\underset{\displaystyle O}{|}}{Si}} - O - \overset{\overset{\displaystyle R}{|}}{\underset{\underset{\displaystyle R}{|}}{Si}} - O - \overset{\overset{\displaystyle O^-}{|}}{\underset{\underset{\displaystyle O}{|}}{Si}} - R$$

Cross-linked silicone

Silicones are used commercially as oils, greases, resins, and synthetic rubbers. The main difference between these products lies in their molecular weight and the type of end and side groups. Silicone oils and silicone greases are linear polymers of low molecular weight; their viscosity increases with increasing molecular weight. Silicone resins are curable compounds which can be either processed to make fiber-reinforced laminates or used as molding compounds to make reinforced parts. Silicone polymer is a silicone of high molecular weight which can be cross-linked to silicone rubber. The curing of resins as well as the cross-linking of silicone polymers requires chemical reactions which increase the size of the molecule; these reactions fall within the scope of the polymer processor. These reactions involve specific side groups (designated as R) or groups which are in end positions. This makes R very important. The organic radicals R are mostly methyl or phenyl groups. In order to modify the properties, such as lowering the brittle point, it is necessary to replace some of the methyl groups. The phenyl group does not show a particularly high reactivity during cross-linking. A certain number of methyl groups are replaced by γ-trifluoropropyl groups to improve resistance to organic solvents (fluorosilicone rubber). Substitution of some of the methyl groups by vinyl groups will make the chemically indifferent silicone more reactive to reactions of an organic nature and will provide polymerizable groups.

Laminated plastics and molding compounds are cured at high temperatures by catalysts such as triethanolamine or a mixture of dibutyltin diacetate and lead 2-ethylhexoate [18]. The molded parts are cured under pressure at 150–175 °C for 30–60 minutes with subsequent afterbake. The chemical reaction of the resin cure is a continuation of the synthesis. The unreacted terminal hydroxyl groups formed during synthesis of the resin will be converted at high temperatures and thereby increase the size of the polymer molecule.

Silicones which contain hydroxyl groups may be cured at room temperature if polyalkyl silicates, for instance, tetramethyl silicate, and a catalyst such as tin octoate or dibutyltin dilaurate are added to the silicone [19, 6]. Polyalkoxysilane acts as the cross-linking agent; it is added before the reaction begins. Water is absolutely necessary, and completely dry resins will not be cured if moisture is totally excluded. The primary step of the reaction is the hydrolysis of the alkoxy groups to silanol groups which are then capable of reacting with the hydroxyl groups of the silicone (condensation).

The rate of polymerization is directly proportional to the concentration of cross-linking agent, the catalyst, temperature, and water content. Acids will inhibit the condensation; which means that volatile acids may be used as temporary inhibitors, and their effectiveness decreases as they volatilize. If water is completely excluded, one-component systems are possible which react only when absorbing by contact with air.

In addition to cross-linking by intramolecular polycondensation of hydroxyl groups, silicones can be cross-linked by reacting hydroxyl groups from silanols with hydrogen groups from silanes [20]. During this reaction, elimination of hydrogen, takes place and the reaction can be classified as a polycondensation.

$$-\overset{|}{\underset{|}{Si}}-O-\overset{|}{\underset{|}{Si}}-OH \quad + \quad H-\overset{|}{\underset{|}{Si}}- \quad \longrightarrow \quad -\overset{|}{\underset{|}{Si}}-O-\overset{|}{\underset{|}{Si}}-O-\overset{|}{\underset{|}{Si}}- \quad + \quad H_2$$

While every one of the previously discussed methods for silicones can be described as a polycondensation, available side chains containing vinyl groups allow cross-linking without the production of decomposition products.

The addition of hydrogen (from the silane compound) onto the vinyl group will result in cross-linking. Platinum, platinum compounds, or peroxide act as catalysts and accelerate the reaction [21]:

$$-\overset{|}{\underset{|}{Si}}-O-\overset{|}{\underset{|}{Si}}-CH=CH_2 \quad + \quad H-\overset{|}{\underset{|}{Si}}- \quad \xrightarrow{Catalyst} \quad -\overset{|}{\underset{|}{Si}}-O-\overset{|}{\underset{|}{Si}}-CH_2-CH_2-\overset{|}{\underset{|}{Si}}-$$

Cross-linking by peroxides is one of the more important methods. Theoretically, the reaction will occur if pure polydimethylsiloxane is used [22], but a certain percentage of vinyl groups will accelerate the reaction considerably [23]. This is the preferred method for vulcanization of silicone rubber. Replacement of only 1% of the methyl side groups by vinyl side groups is sufficient to accelerate the vulcanization. A large number of peroxides may be used. In principle, even peroxides which decompose at low temperatures are suitable, since they can easily be mixed with the polymer at these temperatures. 2,4-Dichlorobenzoyl peroxide is frequently used. During cross-linking by peroxides, primary free radicals remove hydrogen atoms from the methyl group (see Eq. a); in addition, free radicals may be generated by adding primary free radicals to the vinyl groups (see Eq. c). Therefore, several different reaction mechanisms can be used to describe the cross-linking process (see Eqs. b, d):

$$-\overset{|}{\underset{|}{Si}}-O-\overset{\overset{\displaystyle CH_3}{|}}{\underset{|}{Si}}-O-\overset{|}{\underset{|}{Si}}- \quad + \quad R-O\bullet \quad \longrightarrow \quad -\overset{|}{\underset{|}{Si}}-O-\overset{\overset{\displaystyle CH_3}{|}}{\underset{\underset{\displaystyle \bullet}{\displaystyle CH_2}}{Si}}-O-\overset{|}{\underset{|}{Si}}- \quad + \quad R-OH \qquad (a)$$

Primary
free radical

$$2 \; -\overset{|}{\underset{|}{Si}}-O-\overset{\overset{\displaystyle CH_3}{|}}{\underset{\underset{\displaystyle \bullet}{\displaystyle CH_2}}{Si}}-O-\overset{|}{\underset{|}{Si}}- \quad \longrightarrow \quad \begin{array}{c} -\overset{|}{\underset{|}{Si}}-O-\overset{\overset{\displaystyle CH_3}{|}}{\underset{|}{Si}}-O-\overset{|}{\underset{|}{Si}}- \\ | \\ CH_2 \\ | \\ CH_2 \\ | \\ -\overset{|}{\underset{|}{Si}}-O-\overset{|}{\underset{\underset{\displaystyle CH_3}{|}}{Si}}-O-\overset{|}{\underset{|}{Si}}- \end{array} \qquad (b)$$

$$-\overset{|}{\underset{|}{Si}}-O-\overset{\overset{\displaystyle CH_3}{|}}{\underset{\underset{\underset{\displaystyle CH_2}{\|}}{\displaystyle CH}}{Si}}-O-\overset{|}{\underset{|}{Si}}- \quad + \quad R-O\bullet \quad \longrightarrow \quad -\overset{|}{\underset{|}{Si}}-O-\overset{\overset{\displaystyle CH_3}{|}}{\underset{\underset{\displaystyle CH_2-OR}{\bullet CH}}{Si}}-O-\overset{|}{\underset{|}{Si}}- \qquad (c)$$

$$\underset{\substack{\mid \\ \bullet CH \\ \mid \\ CH_2-OR}}{\overset{\substack{CH_3 \\ \mid \\ \mid}}{-Si-O-Si-O-Si-}} \quad + \quad \underset{\substack{\mid \\ CH_2 \\ \bullet}}{\overset{\substack{CH_3 \\ \mid \\ \mid}}{-Si-O-Si-O-Si-}} \quad \longrightarrow \quad \text{(d)}$$

$$\begin{array}{c} \overset{\substack{CH_3 \\ \mid \\ \mid}}{-Si-O-Si-O-Si-} \\ \mid \\ CH_2 \\ \mid \\ CH-CH_2-OR \\ \mid \qquad \mid \qquad \mid \\ -Si-O-Si-O-Si- \\ \mid \qquad \mid \qquad \mid \\ CH_3 \end{array}$$

Silicones which contain vinyl groups may also be vulcanized by sulfur in the presence of sulfur initiators [6]. This type of vulcanization is of interest only for the vulcanization of mixtures containing natural rubber and silicone rubber.

Bibliography to Chapter 10

[1] *Barton, J., J. Pavlinec:* Plaste Kautsch. 15 (1968) 397.
[2] US-PS 3.294.869, Herules Inc.
[3] *Köhnlein, E.:* Kunststoffe 65 (1975) 583.
[4] DE-BP 1.136.485 (1958), Bayer AG.
[5] *Rätzsch, M., E. Braun, M. Gebauer, G. Gladiau:* Plaste Kautsch. 22 (1975) 322.
[6] *Hofmann, W.:* »Vulcanization and Vulcanizing Agents«, MacLaren and Sons, Ltd., London, Palmeton Publishing Co., Inc., New York, S. 251.
[7] *Hofmann, W.:* in: S. Boström »Kautschuk-Handbuch«, Verlag Berliner Union, Bd. 4, 1961, S. 342.
[8] DE-BP 910.221 (1940), Farbenfabriken Bayer.
[9] DE-BP 952.940 (1953), Farbenfabriken Bayer.
[10] DE-BP 968.566 (1954), Farbenfabriken Bayer.
[11] DE-BP 1.071.948 (1958), Farbenfabriken Bayer.
[12] DE-BP 1.054.235 (1955), General Tire & Rubber Co.
[13] *Peter, J.:* in: R. Vieweg; A. Höchtlen »Polyurethane«, Kunststoff-Handbuch, Bd. VII, Carl Hanser Verlag, München, 1966, S. 263.
[14] DE-BP 1.111.379 (1959) Farbenfabriken Bayer.
[15] *Schöneberg, A., A. Mostafa:* J. Chem. Soc. (1943) 654.
[16] *Andersen, D. E., R. G. Arnold:* Ind. Engng. Chem. 45 (1953) 2727.
[17] *Pariser, R.:* Kunststoffe 50 (1960) 623.
[18] *Fordham, S.:* »Silicones«, George Newness Ltd., 1960.
[19] *Hofmann, W.:* Lit. [7], S. 335.
[20] DE-AS 1.058.254, Wacker Chemie.
[21] DE-BP 1.097.133, Wacker Chemie.
[22] *McGregor:* »Silicones and their uses«, McGraw-Hill, 41 (1954), S. 165.
[23] *Noll, W.:* »Chemie u. Technologie der Silicone«, Verlag Chemie, Weinheim, 1960, S. 148.

11 Cross-linking of Ethylenic Unsaturated Polymers Using Sulfur

11.1 Ethylenic Unsaturated Elastomers

Diene rubbers which can be cross-linked by sulfur are summarized in Table 49. Nitrile rubber is a copolymer consisting of acrylonitrile and butadiene, with a content of acrylonitrile between 18 and 48%. Pure polyisobutylene cannot be cross-linked by sulfur or peroxides. Butyl rubber is not a pure polyisobutylene, it is a copolymer containing 1–4% isoprene. The isoprene introduces double bonds into the polymer chain, thereby permitting vulcanization by sulfur. Another rubber which contains double bonds is polychloroprene. Although, in principle, it can be cross-linked by sulfur, better results are obtained if ZnO + MgO, or substituted thioureas, are used as vulcanization agents (see section 10.7).

Table 49   Diene rubbers which can be vulcanized by sulfur

	Abbreviation	stylized formula
Natural rubber	NR	$(-CH_2-\overset{\overset{\displaystyle CH_3}{\vert}}{C}=CH-CH_2-)_n$
Isoprene rubber	IR	same as natural rubber
Styrene-butadiene rubber	SBR	$(-CH_2-\overset{\vert}{CH}-)_n -(CH_2-CH=CH-CH_2-)_m$ 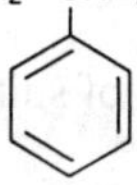
Polybutadiene rubber	BR	$(-CH_2-CH=CH-CH_2-)_n$
Nitrile rubber	NBR	$(-CH_2-\overset{\overset{\displaystyle}{\vert}}{CH}-)_n -(CH_2-CH=CH-CH_2-)_m$ CN
Butyl rubber	IIR	$(-CH_2-\overset{\overset{\displaystyle CH_3}{\vert}}{\underset{\underset{\displaystyle CH_3}{\vert}}{C}}-)_n -(CH_2-\overset{\overset{\displaystyle CH_3}{\vert}}{C}=CH-CH_2-)_m$

All the polymers mentioned so far represent elastomers which do not have any practical applications as such but can be converted to a useful material by vulcanization. If the amount of sulfur used is about 5% or less, the vulcanized material exhibits rubberlike elastic properties and is therefore not considered a plastic material according to customary terminology. In terms of chemistry, however, there is no basis for the distinction between a plastic and an elastomer. The distinction is based merely on the different mechanical properties of the materials.

11.2   Principles of Vulcanization with Sulfur

If a mixture of natural rubber and sulfur is heated for an extended period of time, some of the sulfur is inserted between polymer chains, resulting in the formation of cross-linking sites, while the rest of the sulfur produces intramolecular cross-links or causes ring formation. Sulfur acts as a bifunctional reactant which is able to react with ethylenic unsaturated polymer molecules.

The vulcanization of natural rubber, or other elastomers using only sulfur proceeds very slowly and results in products with poor properties. The fact that natural rubber can be cross-linked at all with sulfur (without any accelerator) is made possible by the contaminants present in natural rubber. If natural rubber is extracted by acetone, vulcanization of the purified polymer by pure sulfur is almost impossible.

The commercial vulcanization with sulfur requires special additives which accelerate the process. The accelerated vulcanization results in products which have better properties than those which are cross-linked using pure sulfur. At ordinary temperature, pure sulfur is a cross-linking agent of very low reactivity, which increases only above $160\,°C$. There is, however, no better cross-linking agent. The sulfur homologs selenium and tellurium may also be used as cross-linking agents; however, their use is justified only in special applications because of their higher price. The low reactivity of sulfur is stipulated by its ring structure. At room temperature, sulfur exists in the form of yellow rhombic crystals, which consist of cyclic S_8 molecules.

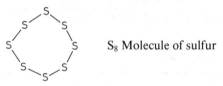 S_8 Molecule of sulfur

If sulfur is heated to $95.6\,°C$ it changes to the yellow, monoclinic form, and then at $119\,°C$ it transforms into a yellow, low-viscosity melt. Up to about $160\,°C$ both the solid and the melt consist of the cyclic S_8 molecules. Thermal cleavage of the cyclic structure occurs only above this temperature; the resulting biradicals polymerize, increasing the melt viscosity. Only as a result of thermal cracking at substantially higher temperatures will the molecular weight decrease. At $444.6\,°C$, the sulfur begins to boil; it exists at this temperature in the form of S_6 molecules, which on further heating break down into even smaller molecules.

Since the cyclic S_8 molecule is unable to react with rubber, the cyclic structure must first be converted to a more active form. Attempts to cross-link rubber with sulfur at approximately $160\,°C$ will produce cross-linking due to the ring cleavage at this temperature, but the cross-linking sites are far apart and large amounts of sulfur are needed for the vulcanization. The resulting product may be described schematically as follows:

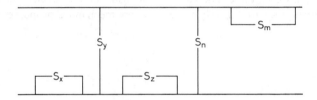

Optimum cross-linking methods will result in limited intramolecular cross-linking and, in particular, in short sulfur chains between cross-linking sites. Cross-links consisting of only single sulfur atoms result in rubber having attractive properties. To some extent, even better properties can be achieved if the chains are linked without sulfur bridges. The higher bond energies are responsible for the better properties.

$-C-S_x-C-$ bond energy <268 kJ/mole

$-C-S-S-C-$ bond energy 268 kJ/mole

$-C-S-C-$ bond energy 285 kJ/mole

$-C-C-$ bond energy 352 kJ/mole

The simple C-C linkage, therefore, will result in products with the best thermal stability.

Cross-linking with sulfur is a dynamic process, in which the polysulfide cross-linking sites themselves can again participate in the process of vulcanization. The probability of cleavage at an existing cross-linking site increases with the length of the chain, in other words, with the number of sulfur atoms in that chain. Therefore, very short sulfur chains are more stable.

The activators used in commercial vulcanization with sulfur react with sulfur to form intermediates which make the sulfur available in the form of an open chain of low molecular weight. The effect of the activators is twofold: they lower the activation energy of the cross-linking reaction, and they cleave the S_8 ring into suitable fragments.

The large number of types of rubber and the multitude of vulcanization conditions necessary for cross-linking require a very large number of cross-linking additives.

11.3   Chemical Reactions during the Vulcanization of Diene Polymers

The reactivity of the diene polymers is due to their double bonds and the ease with which the hydrogen atoms located in the position alpha to the double bonds may be abstracted. The double bonds can undergo either ionic or free radical addition reactions, e.g.

$$-CH_2-CH=CH-CH_2-$$

Segment of the
polybutadiene chain

$$-CH_2-\overset{-}{C}H-\overset{+}{C}H-CH_2- \quad\quad (a)$$

$$-CH_2-\overset{\cdot}{C}H-\overset{\cdot}{C}H-CH_2- \quad\quad (b)$$

$$-\underset{\underset{H}{\cdot}}{C}H-CH=CH-CH_2- \quad\quad (c)$$

The reactive behavior of the diene polymer depends on the nature of the reactant. If the reactant is ionic, the polymer reacts according to (a); if it is a free radical, the diene polymer reacts according to (b) and (c), with c) being of particular importance. While the vulcanization depends on the presence of ethylenic unsaturated groups, it is not necessarily accompanied by a decrease in the concentration of these groups.

Sulfur can also react either ionically or as a free radical. The strictly thermal cleavage of S_8 rings yields diradicals. The reaction with ionic components results in polarization of the sulfur molecules.

$$\cdot S-S-S-S-S-S-S-S\cdot \qquad \text{Radical cleavage}$$

$$S_8$$

$$^-S-S-S-S-S-S-S-S^+ \qquad \text{Ionic cleavage}$$

The chemical mechanism of the sulfur vulcanization with or without accelerators has been described by several theories, which are summarized, for instance, in [1]. It is possible that all these theories have some validity but to different degrees. The essential reactions during cross-linking are (using polyisoprene as an example):

$$(-CH_2-CH=C-CH_2-) \;+\; :S_x \;\longrightarrow\; (-CH_2-CH=C-\overset{\cdot}{C}H-) \;+\; \cdot S_x H \qquad (a)$$
$$\overset{|}{CH_3} \qquad\qquad\qquad\qquad\qquad\qquad \overset{|}{CH_3}$$

$$(-CH_2-CH=C-\overset{\cdot}{C}H-) \;+\; S_8 \;\longrightarrow\; (-CH_2-CH=C-CH-) \;+\; \cdot S_{8-x} \qquad (b)$$
$$\overset{|}{CH_3} \qquad\qquad\qquad\qquad\qquad\qquad \overset{|}{\underset{\cdot S_x}{CH-}}$$

$$(-CH_2-CH=\overset{\overset{CH_3}{|}}{C}-CH-) \;+\; (-CH_2-CH=\overset{\overset{CH_3}{|}}{C}-CH_2-) \;\longrightarrow \qquad (c)$$
$$\overset{|}{\cdot S_x}$$

$$\longrightarrow \quad (-CH_2-CH=\overset{\overset{CH_3}{|}}{C}-CH-)$$
$$\overset{|}{\underset{(-CH_2-\underset{\cdot}{CH}-\overset{\overset{CH_3}{|}}{C}-CH_2-)}{S_x}}$$

$$(-CH_2-CH=\overset{\overset{CH_3}{|}}{C}-CH-) \;+\; (-CH_2-CH=\overset{\overset{CH_3}{|}}{C}-CH_2-) \;\longrightarrow \qquad (d)$$
$$\overset{|}{\underset{(-CH_2-CH-\overset{\cdot}{C}-CH_2-)}{S_x}}$$
$$\overset{|}{CH_3}$$

$$\longrightarrow \quad (-CH_2-CH=\overset{\overset{CH_3}{|}}{C}-CH-) \;+\; (-CH_2-CH=\overset{\overset{CH_3}{|}}{C}-\overset{\cdot}{C}H-)$$
$$\overset{|}{\underset{(-CH_2-CH-CH-CH_2-)}{S_x}}$$
$$\overset{|}{CH_3}$$

Therefore, the strictly thermal sulfur cross-linking with sulfur is predominantly a free radical chain reaction. The SH radical formed during the chain initiation (Eq. a) is capable of either producing chain radicals, in which case thiohydrogen sulfide is formed, or of adding to a chain. Since the action of S_8 results in the formation of reactive. S fragments, the distance between the cross-link sites is variable.

Since the activation energy necessary for the cleavage of S_8 molecules is approximately equal to that necessary for the cleavage of long chains of sulfur linkages, these linkages can be cleaved again to form new bonds:

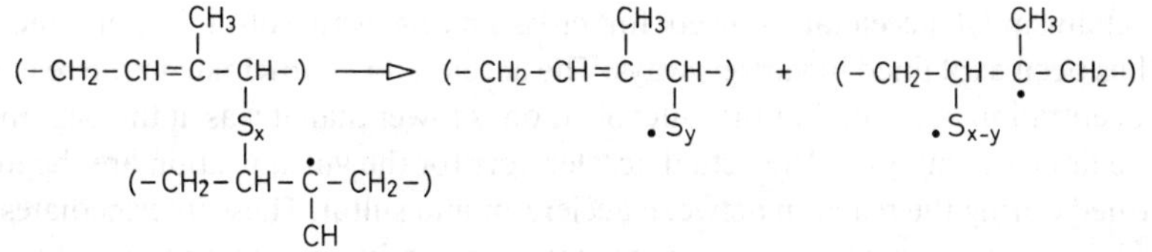

$$(-CH_2-CH=\overset{\overset{\displaystyle CH_3}{|}}{C}-CH-) \longrightarrow (-CH_2-CH=\overset{\overset{\displaystyle CH_3}{|}}{C}-CH-) + (-CH_2-CH-\overset{\overset{\displaystyle CH_3}{|}}{\underset{\displaystyle \bullet}{C}}-CH_2-)$$

with the S_x, •S_y, •S_{x-y} groups and

$$(-CH_2-\overset{|}{CH}-\overset{\overset{\displaystyle \bullet}{}}{C}-CH_2-)$$
$$\underset{\displaystyle CH}{|}$$

11.4 Activators for Cross-linking with Sulfur

A large number of compounds are used as accelerators for cross-linking with sulfur. The most important classes of accelerators are summarized in Table 50. Other accelerating components are zinc oxide (ZnO) and stearic acid, which are present in most rubber recipes.

Tabelle 50 Accelerators for cross-linking with sulfur

Formula	Name	Application (example)
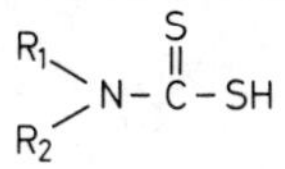 R_1, R_2 N–C(=S)–SH	Dithiocarbamic acid	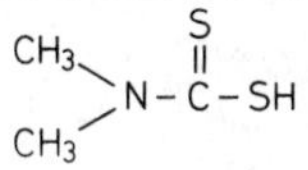 CH_3, CH_3 N–C(=S)–SH
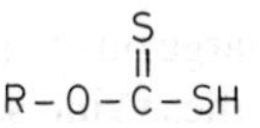 R–O–C(=S)–SH	Xanthogenic acid	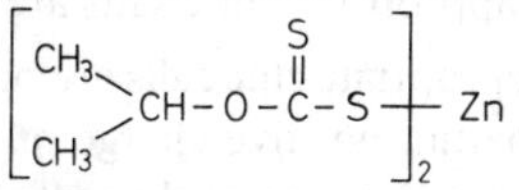 $[CH_3, CH_3$ CH–O–C(=S)–S$]_2$ Zn
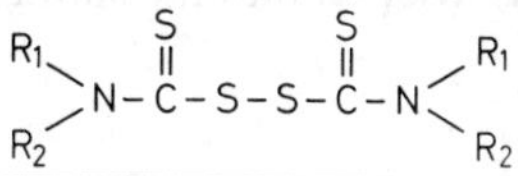 R_1, R_2 N–C(=S)–S–S–C(=S)–N R_1, R_2	Thiuramdisulfide	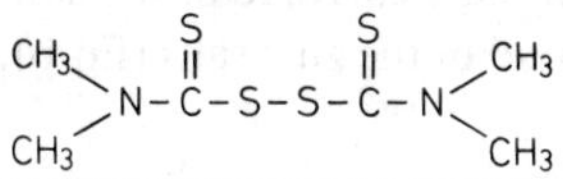 CH_3, CH_3 N–C(=S)–S–S–C(=S)–N CH_3, CH_3
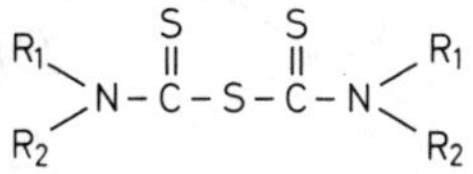 R_1, R_2 N–C(=S)–S–C(=S)–N R_1, R_2	Thiurammonosulfide	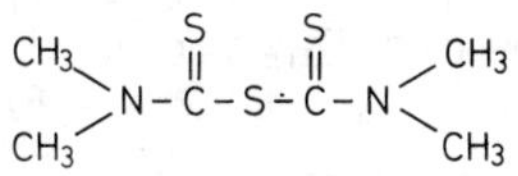 CH_3, CH_3 N–C(=S)–S⋅C(=S)–N CH_3, CH_3
R, R |_N=C–SH (thiazole ring)	Mercaptothiazole	benzothiazole–C–SH
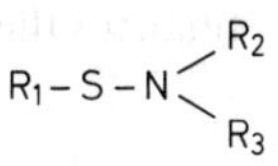 R_1–S–N R_2, R_3	Sulfenamide	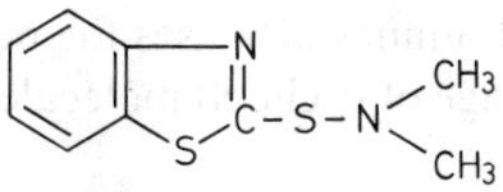 benzothiazole–C–S–N CH_3, CH_3
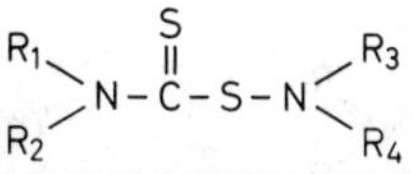 R_1, R_2 N–C(=S)–S–N R_3, R_4	Tetraalkyldithio-carbamylsulfenamide	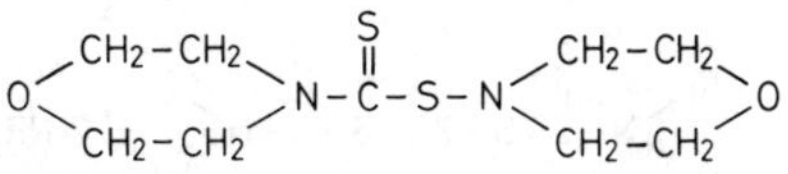 O(CH_2–CH_2, CH_2–CH_2)N–C(=S)–S–N(CH_2–CH_2, CH_2–CH_2)O
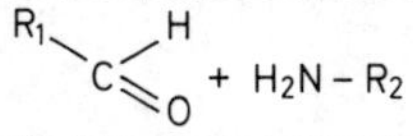 R_1, H C=O + $H_2N–R_2$	Aldehyde amine	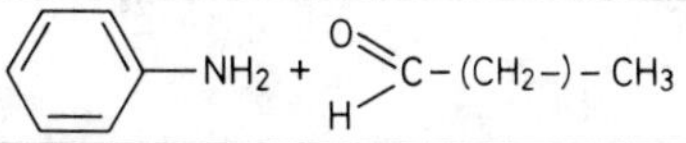 phenyl–NH_2 + O=C(H)–(CH_2–)–CH_3
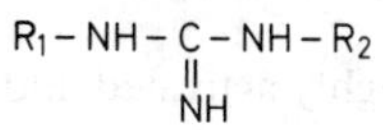 R_1–NH–C(=NH)–NH–R_2	Diarylguanidine	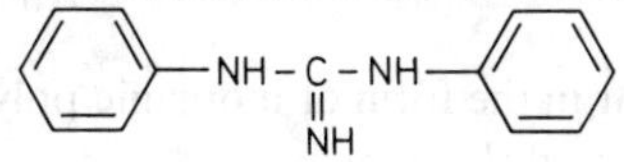 phenyl–NH–C(=NH)–NH–phenyl

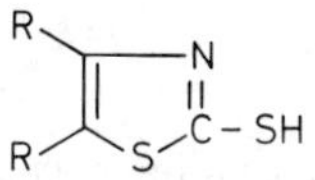

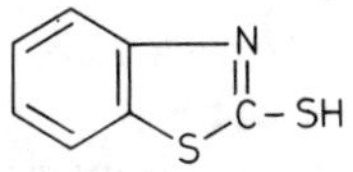

The mechanism of accelerators used for cross-linking with sulfur is not precisely known and has been explained in various ways. The accelerators participate in the reaction, and their concentration at the end of the vulcanization is lower than it was at the beginning. Thus, they are not true catalysts. The actual accelerators for the vulcanization are the intermediates formed during the reaction between acclerator and sulfur. These intermediates decompose, and the starting components are regenerated; in addition to regenerating the activator, the intermediates yield several other components.

Thiocarbamic acids and xanthic acids are very similar in their behavior as accelerators. Their effect as accelerators on systems which contain only sulfur and rubber is very limited. As with many other accelerators, it is necessary to employ a system consisting of several components in order to achieve in acceleration [1]. The effect of thiocarbamic acids and xanthic acids is enhanced by the use of ZnO as a secondary accelerator. Stearic acid causes further activation of the accelerator system.

In addition to other possibilities of reaction, the acidic accelerator may be neutralized to form zinc salts (Eq. a):

$$
2 \quad \underset{R_2}{\overset{R_1}{>}}N-\overset{\overset{S}{\|}}{C}-SH \; + \; ZnO \; \longrightarrow \; \underset{R_2}{\overset{R_1}{>}}N-C\overset{S}{\underset{S}{<}} \; Zn \; \overset{S}{\underset{S}{>}}C-N\underset{R_2}{\overset{R_1}{<}} \; + \; H_2O \tag{a}
$$

Dithiocarbamic   Zinc dithiocarbamate
acid

Therefore, appropriate zinc salts are also suitable accelerators.

The appropriate zinc salts are polarized by forming a partial negative charge on the sulfur and a partial positive charge on the metal atom. It is conceivable that these zinc salts cause an ionic cleavage of the sulfur and the open-chain sulfur is inserted into the sulfur-metal bond of the zinc salts (Eq. b):

$$
\underset{R_2}{\overset{R_1}{>}}N-\overset{\overset{S}{\|}}{C}-S^--\overset{2+}{Zn}-S^--\overset{\overset{S}{\|}}{C}-N\underset{R_2}{\overset{R_1}{<}} \quad \longrightarrow \quad \underset{R_2}{\overset{R_1}{>}}N-\overset{\overset{S}{\|}}{C}-S-S_8-Zn-S-\overset{\overset{S}{\|}}{C}-N\underset{R_2}{\overset{R_1}{<}} \tag{b}
$$

The accelerator effect is increased by the use of amines as secondary accelerators. The addition of amines increases the polarization of the zinc salts, which further enhances the ionic cleavage of cyclic S_8 molecules (Eq. c):

$$
\underset{R_2}{\overset{R_1}{>}}N-\overset{\overset{S}{\|}}{C}-S-Zn-S-\overset{\overset{S}{\|}}{C}-N\underset{R_2}{\overset{R_1}{<}} \; + \; 2\,NR_3 \; \longrightarrow \; \underset{R_2}{\overset{R_1}{>}}N-\overset{\overset{S}{\|}}{C}-S-\overset{\overset{NRRR}{|}}{\underset{\underset{NRRR}{|}}{Zn}}-S-\overset{\overset{S}{\|}}{C}-N\underset{R_2}{\overset{R_1}{<}} \tag{c}
$$

Amine   Complex compound

Sulfur in the form of inorganic polysulfide, as in the zinc salts, is highly activated and easily released (Eq. d):

$$\begin{array}{c} R_1 \\ \diagdown \\ R_2 \end{array} N-\overset{\overset{\textstyle S}{\|}}{C}-S-S_8-Zn-S-\overset{\overset{\textstyle S}{\|}}{C}-N \begin{array}{c} R_1 \\ \diagup \\ R_2 \end{array} \longrightarrow \ \overset{\bullet}{S_x} + \begin{array}{c} R_1 \\ \diagdown \\ R_2 \end{array} N-\overset{\overset{\textstyle S}{\|}}{C}-S-S_{8-x}-Zn-S-\overset{\overset{\textstyle S}{\|}}{C}-N \begin{array}{c} R_1 \\ \diagup \\ R_2 \end{array} \qquad (d)$$

If the sulfur is released in the form of very small molecules ($X = 1$ or 2), the more stable short-chain sulfur linkages are formed, which are less likely to undergo changes during further vulcanization.

The accelerating effect of thiuram disulfides is, in part, due to a free radical mechanism. The thermal decomposition of thiuram disulfide may result in the formation of free radicals (Eq. e) which may cause the radical cleavage of cyclic S_8 molecules (Eq. f) yielding short chain sulfur molecules (Eq. g):

$$\begin{array}{c} R_1 \\ \diagdown \\ R_2 \end{array} N-\overset{\overset{\textstyle S}{\|}}{C}-S-S-\overset{\overset{\textstyle S}{\|}}{C}-N \begin{array}{c} R_1 \\ \diagup \\ R_2 \end{array} \longrightarrow \ 2 \ \begin{array}{c} R_1 \\ \diagdown \\ R_2 \end{array} N-\overset{\overset{\textstyle S}{\|}}{C}-S^{\bullet} \qquad (e)$$

$$\begin{array}{c} R_1 \\ \diagdown \\ R_2 \end{array} N-\overset{\overset{\textstyle S}{\|}}{C}-S^{\bullet} + S_8 \longrightarrow \begin{array}{c} R_1 \\ \diagdown \\ R_2 \end{array} N-\overset{\overset{\textstyle S}{\|}}{C}-S-S_8^{\bullet} \qquad (f)$$

$$\begin{array}{c} R_1 \\ \diagdown \\ R_2 \end{array} N-\overset{\overset{\textstyle S}{\|}}{C}-S-S_8^{\bullet} \longrightarrow \begin{array}{c} R_1 \\ \diagdown \\ R_2 \end{array} N-\overset{\overset{\textstyle S}{\|}}{C}-S-S_{8-x}^{\bullet} + \overset{\bullet}{S_x} \qquad (g)$$

This mechanism is supported by the fact that the cross-linking of diene rubbers with tetramethylthiuram disulfide is possible in the absence of sulfur. If thiuram monosulfide is used together with an excess of sulfur, thiuram disulfide is presumably formed as an intermediate.

It is believed that 2-mercaptobenzothiazole, the most important representative of the mercaptothiazoles, reacts primarily with ZnO, thus forming zinc salts (Eq. h):

$$2 \ \underset{S}{\overset{N}{\diagdown}}C-S-H \ \xrightarrow{\ + ZnO\ } \ \underset{S}{\overset{N}{\diagdown}}C-S-Zn-S-C\underset{S}{\overset{N}{\diagup}} \qquad (h)$$

2-Mercaptobenzo-
thiazole

These zinc salts add on sulfur similarly to the zinc dithiocarbamates discussed before (Eq. i):

$$\underset{S}{\overset{N}{\diagdown}}C-S-Zn-S-C\underset{S}{\overset{N}{\diagup}} \ + \ S_8 \ \longrightarrow \qquad (i)$$

$$\underset{S}{\overset{N}{\diagdown}}C-S-S_8-Zn-S-C\underset{S}{\overset{N}{\diagup}}$$

The actual cross-linking agents are generated by elimination of small sulfur diradicals.

2-Mercaptobenzothiazole is an effective accelerator only in the presence of long-chain fatty acids, such as stearic acid. The zinc salts of 2-mercaptobenzothiazole are only slightly soluble in rubber and therefore do not exhibit their full effectiveness. Stearic acid forms addition complexes with the accelerator, thus increasing the solubility of the accelerator in the rubber, which in turn enhances its effectiveness.

During vulcanization, sulfenamides based on 2-mercaptobenzothiazole initially cleave into free radicals (Eq. j), which are converted by radical transfer into 2-mercaptobenzothiazole and an amine (Eq. k) [1]. Both products propagate the cross-linking reaction:

(j)

(k)

2-Mercaptobenzothiazole

Dimethyl
amine

The free radicals formed initially may cause the hydrogen to be eliminated directly from the diene polymer. The newly formed 2-mercaptobenzothiazole reacts with ZnO as described above, yielding zinc salts, which add on sulfur. According to the mechanism discussed before, the amine, formed likewise, can activate the zinc salts thereby, accelerating the insertion of sulfur.

Alkyl dithiocarbamyl sulfenamides are structurally similar to mercaptothiazyl sulfenamides. The difference is substitution of the mercaptothiazyl group by the dithiocarbamyl group, which can also accelerate the vulcanization. Therefore, this accelerator behaves like a sulfenamide and like a dithiocarbamic acid accelerator.

The cross-linking of rubber occurs predominantly by means of free radical mechanism. Those hydrogen atoms which are adjacent to the chain double bond react preferentially. They are abstracted partially, and the resulting polymer radicals react with sulfur or sulfur-containing compounds to form cross-linking sites. These cross-linkages consist of X sulfur atoms.

The use of sulfur can be completely eliminated if tetramethylthiuram disulfide is used as an accelerator. At elevated temperatures, cleavage of the accelerator occurs and the generated radicals react with the rubber to yield cross-linkable polymer radicals. The cross-linked polymer does contain sulfur chains, since the accelerator also acts as a sulfur donor. If the radical formation occurs in the presence of sulfur, the sulfur is incorporated according to a purely radical mechanism.

Besides free radical reactions, the cross-linking of rubber also involves addition reactions of -S$_x$-SH or -SH groups to double bonds.

In the mixture, the ratio of sulfur to rubber depends on several factors. As a first approximation, increasing the amount of sulfur results in a higher degree of vulcanization. However, since the degree of vulcanization depends on the length of the sulfur chains at the linkage points, in some cases smaller amounts of sulfur will result in improved degrees of vulcanization depending on the effectiveness of the accelerator.

In addition, diene rubbers can also be cross-linked with peroxides, which link the polymer chains by C-C bonds directly. Because of the higher thermal stability of the C-C bond than of the C-S and S-S bonds, elastomers cured by peroxides are particularly suitable for molded parts intended for use at high temperature.

The chemical reaction mechanism of the peroxide cross-linking is similar to the cross-linking mechanism of polyethylene. Unstable hydrogen atoms which are alpha to the double bond are cleaved by primary free radicals (Eq. m). The newly formed chain radicals formed in this manner dimerize, resulting in cross-linkage (Eq. n):

$$R-O-O-R \longrightarrow 2\ R-O^{\bullet} \tag{l}$$

Peroxide   Initial radical

$$-CH_2-\underset{\underset{CH_3}{|}}{C}=CH-CH_2- \ +\ R-O^{\bullet} \longrightarrow -\overset{\bullet}{C}H-\underset{\underset{CH_3}{|}}{C}=CH-CH_2- \ +\ R-OH \tag{m}$$

$$2\ -\overset{\bullet}{C}H-\underset{\underset{CH_3}{|}}{C}=CH-CH_2- \longrightarrow \begin{array}{c} CH_3 \\ | \\ -CH-C=CH-CH_2- \\ -CH-C=CH-CH_2- \\ | \\ CH_3 \end{array} \tag{n}$$

Bibliography to Chapter 11

[1] *Hofmann, W.:* »Vulcanisation and Vulcanizing Agents«, MacLaren and Sons, Ltd., London; Palmeton Publishing Co., New York.

12   Degradation of Polymers during Processing

The chemical reactions discussed so far, are carried out intentionally to produce or modify synthetic polymers in the processing plant. It is characteristic of these reactions that the molecular weight of the starting materials will be increased. In addition to those processes in which a chemical synthesis is conducted intentionally, several processes exist during which undesired reactions occur which result in a material with reduced molecular weight.

During the thermoplastic processing of various plastic materials, decomposition of macromolecules occurs. These unintended reactions may have a detrimental effect on the quality of the material. Therefore, extensive degradation of the polymer must be avoided. The plastic processor should try to prevent these reactions or at least minimize their detrimental effects as much as possible.

The degradation during processing of synthetic polymers is caused by

(a)   Thermal chain cleavage
(b)   Oxidative decomposition
(c)   Hydrolytic degradation

The melt processing of some polymers requires temperatures at which the material already starts to decompose. During processing, friction energy is dissipated to the hot material, which may result in hot spots or in mechanically induced chain cleavage. Oxidative degradation cannot be avoided completely, because air dissolves in the material and is therefore introduced into the hot melt. Hydrolytic polymer degradation is of importance only in the case of those polymers which can be hydrolyzed, such as polyesters and polyamides. Since water gets into the polymer at least in trace quantities, its effect must always be considered.

In principle, there are two ways to control deleterious degradation:

(a)   Through the introduction of stabilizers
(b)   By avoiding unnecessary thermal exposure of the polymer and by excluding oxygen and water as much as possible.

Most plastics are compounded by the manufacturer with a sufficient amount of one or more suitable stabilizers. By maintaining optimum processing conditions, the plastic processor must then limit the degradation to a minimum. In the processing of PVC, the stabilizers are frequently added to the polymer by the processor himself. Thus, the processor can select a stabilizer system which is suitable for his method of processing and the intended application.

During the processing of the polymer, the stabilizers react with the intermediate compounds that were formed as a result of the polymer degradation. During the processing, therefore, stabilizers are consumed and must be added in sufficient quantities.

The catalytic effect of traces of metals, which are released due to the abrasive effect of the polymer melt on the surfaces of the processing equipment, is an additional cause of polymer degradation. The effect of the metals can be eliminated by using suitable stabilizers to form complex compounds. However, the processor has to select adequately protected processing equipment for the processing of particularly abrasive compositions.

The chemical reactions which occur during the degradation of polymers vary from polymer to polymer. A detailed discussion of the stabilization of polymers against light and heat is provided in [1]. Discussion of degradation reactions of some important polymers, as well as methods to control these reactions, follows.

12.1 Degradation Reactions of Polyvinyl Chloride

With time, unstabilized polyvinyl chloride (PVC), if exposed to temperatures of more than 100 °C wiil turn light yellow, then brown, and finally black. The rate at which this formation and deepening of color occurs increases with increasing temperature. The extrusion and calendering of rigid PVC requires temperatures of 170 °C and higher, and the processing of unstabilized material would not be possible at those temperatures. Initially, the discoloration of PVC is not accompanied by significant deterioration of its mechanical properties. These deteriorate only after the PVC has been damaged to a larger extent.

The discoloration is caused by chemical reactions which start with the elimination of HCl from the PVC molecule [2]. Initially, this elimination of HCl occurs only at certain points, where the chlorine atom is activated due to a unique configuration of the molecule. Reactive positions are introduced automatically into the polymer during polymerization; they can be also formed by oxidation. Chain end groups, such as the acyl groups of peroxides, increase the reactivity of the chain ends. Adjacent chlorine atoms, whose positioning is due to chain termination by the dimerization of two radicals, are also reactive sites. Chain termination which is caused by disproportionation will result in reactive structures at the chain end. Branching sites are occupied by a very reactive tertiary chlorine atom.

$$- CH_2 - CH - CH_2 - CH - O - \overset{\overset{\displaystyle O}{\|}}{C} - R \qquad \text{Reactive acyl end group}$$
$$| |$$
$$Cl Cl$$

Initiator
group

$$- CH_2 - CH - CH_2 - CH - CH - CH_2 - CH - CH_2 - \qquad \text{Reactive chain segment}$$
$$| | | |$$
$$Cl Cl Cl Cl$$

Adjacent
Cl atoms

$$- CH_2 - CH - CH = CH$$
$$| |$$
$$Cl Cl$$

$$- CH_2 - CH - CH_2 - CH_2 - Cl \qquad\qquad\qquad \text{Reactive chain ends}$$
$$|$$
$$Cl$$

$$- CH - CH_2 - C = CH_2$$
$$| |$$
$$Cl Cl$$

$$
\begin{array}{c}
| \\
CHCl \\
| \\
CH_2 \\
| \\
- CH_2 - CH - CH_2 - C - CH_2 - CH - \\
\quad\quad\ | \quad\quad\quad | \quad\quad\quad | \\
\quad\quad Cl \quad\quad\ Cl \quad\quad Cl
\end{array}
\qquad \text{Reactive branch point}
$$

The reactive configuration facilitates the elimination of HCl. Starting at a point where one HCl molecule has already been eliminated, release of successive HCl molecules continues in an unzipping reaction. This results in the formation of polyene structures in which the double bonds are in conjugated positions:

$$
\begin{array}{c}
| \\
CH_2 \\
| \\
- C - CH_2 - CH - CH_2 - CH - CH_2 - CH - CH_2 - \\
| \quad\quad\quad | \quad\quad\quad | \quad\quad\quad | \\
Cl \quad\quad Cl \quad\quad Cl \quad\quad Cl
\end{array}
\ \xrightarrow[-HCl]{}\
\begin{array}{c}
| \\
CH_2 \\
| \\
- C = CH - CH - CH_2 - CH - CH_2 - CH - CH_2 - \\
\quad\quad\quad\ | \quad\quad\quad\ | \quad\quad\quad\ | \\
\quad\quad\quad Cl \quad\quad\ Cl \quad\quad\ Cl
\end{array}
$$

$$
\xrightarrow[-HCl]{}\
\begin{array}{c}
| \\
CH_2 \\
| \\
- C = CH - CH = CH - CH - CH_2 - CH - CH_2 - \\
\quad\quad\quad\quad\quad\quad\ | \quad\quad\quad\ | \\
\quad\quad\quad\quad\quad\ Cl \quad\quad\ Cl
\end{array}
$$

$$
\xrightarrow[-HCl]{}\
\begin{array}{c}
| \\
CH_2 \\
| \\
- C = CH - CH = CH - CH = CH - CH - CH_2 \\
\quad\quad\quad\quad\quad\quad\quad\quad\quad\quad\ | \\
\quad\quad\quad\quad\quad\quad\quad\quad\quad\ Cl
\end{array}
\ \xrightarrow[-HCl]{}\ \text{etc.}
$$

If the conjugated system contains 5–7 or more conjugated double bonds, it will absorb light, resulting in discoloration of the material. Longer sequences of polyenes may undergo intra- or intermolecular cyclization. Initially, the degradation of PVC is not accompanied by chain cleavage. This explains the delayed decrease in strength.

In addition to the reactive sites which are brought into the PVC molecule during the polymerization, reactive sites formed through oxidation of the polymer molecule are of particular importance. Degradation in the presence of oxygen results in additional reactive sites being formed by oxidation and therefore differs from dehydrochlorination in a vacuum or in a nitrogen atmosphere. Consequently, the elimination of HCl is promoted by the presence of oxygen. PVC which has been exposed to UV light in the presence of air is partially oxidized and during after-annealing will release HCl faster than untreated PVC [4].

The ketoallyl group is formed by oxidation

$$
\begin{array}{c}
O \\
\| \\
- C - CH = CH -
\end{array}
$$

and facilitates further HCl elimination nearby. The keto group becomes part of the chromophoric system and therefore accelerates the discoloration. Carbonyl groups can be formed as a result of the oxidizing action of the peroxides during the polymerization or by air oxidation during drying and processing.

On the one hand, air causes increased HCl elimination; on the other hand, it can retard the discoloration of PVC and may even lighten the color of slightly discolored PVC [5]. Oxygen reacts with the polyene groups which cause the discoloration and thereby interrupts the conjugated system, for instance by chain cleavage. The shortening of the polyene sequence results in a lighter color.

In addition to the HCl elimination which is favored at reactive sites, an occasional random release of HCl occurs at ideal chain segments. However, this random release requires a higher energy level [6].

Radicals, which can be formed by the decomposition of peroxides, for instance, remove hydrogen from the PVC molecule. The resulting chain radicals also cause a zipperlike dehydrochlorination. They can also lead to cross-linking of the macromolecules [1 a].

12.2   Stabilization of Polyvinyl Chloride

The processing of PVC requires stabilizers which combine with the generated HCl, prevent the formation of chromophoric groups, and, if possible, retard the elimination of HCL. Table 51 shows stabilizers which are used in the processing of PVC. The mechanisms by which the individual stabilizers act vary. Usually, they are not used by themselves but in combination with other stabilizers. Such combinations of different stabilizers frequently exhibit a synergistic effect.

Tabelle 51   Heat stabilizers for PVC

Formula	Name
$3\ PbO \cdot PbSO_4 \cdot H_2O$	Tribasic lead sulphate
PbO	Lead oxide
BaO / CdO	Barium-cadmium oxide
CaO / ZnO	Calcium-zinc oxide
$Me^{++} \left(-O - \overset{O}{\underset{\parallel}{C}} - C_{17}H_{35} \right)_2$	Ba, Cd, Zn, Ca, Pb-salts of stearic acid
Dibutyltin dilaurate structure	Dibutyltin dilaurate (ester-tin-compound)
Dimethyltin-bis(n-octyl-maleate) structure	Dimethyltin-bis(n-octyl-maleate)
Diglycidylbisphenol-A structure	Diglycidylbisphenol-A
Bis-(diphenylphosphite) of bisphenol-A structure	Bis-(diphenylphosphite) of bisphenol-A

Sodium carbonate is used as the basic stabilizer. It does not change the rate of HCl elimination but will neutralize the toxic and highly corrosive HCl:

$$Na_2CO_3 + 2\ HCl \longrightarrow 2\ NaCl + CO_2 + H_2O$$

Likewise, metal oxides act strictly as HCl scavengers. Basic heavy metal salts, such as basic lead sulfate, serve the same purpose:

$$3\ PbO \cdot PbSO_4 \cdot H_2O + 6\ HCl \longrightarrow 3\ PbCl_2 + 3\ PbSO_4 + 6\ H_2O$$

Similarly, the Ba, Cd, Zn, Ca, and Pb salts of organic carboxylic acids, such as hexanoic, octanoic, lauric, stearic, ricinoleic, maleic, salicylic, phthalic, and naphthenic acid, can, in principle, serve as HCl scavengers.

$$(C_{11}H_{23}COO)_2\ Cd + 2\ HCl \longrightarrow CdCl_2 + 2\ C_{11}H_{23}COOH$$

Cd laurate

Free stearic acid does not affect the HCl elimination. The various metal salts exhibit different effects. The metal chlorides formed by the reaction with HCl do not act to inhibit HCl elimination. On the contrary, they may even accelerate the polymer degradation catalytically [7]. Although stabilizers based on metals do not have a significant effect on the rate of HCl elimination, they do delay the onset of discoloration.

Lead salts are the most effective HCl scavengers. For environmental reasons, however, they are being replaced by other metal salts. Stabilizer combinations used in large quantities are Ba-Cd salts with successive replacement by Ca-Zn stabilizers.

Cadmium stearate retards the discoloration even though the cadmium chloride generated is a catalyst for further degradation. If cadmium stearate is used in combination with barium stearate, the released HCl forms barium chloride, thus prolonging the effect of the cadmium stearate.

The stabilizers discussed so far do not affect the degree of degradation; they mainly prevent the detrimental effect of the degradation products and delay the onset of discoloration. The causes of polymer degradation are not affected by these catalysts. If the active sites of the PVC molecule, where elimination of HCl starts, can be transformed to more stable structures, the causes of polymer degradation are eliminated. Organic tin compounds are particularly useful in causing such reactions. These stabilizers are expensive, however, and for this reason are presently not used in large quantities. Their effectiveness at low concentrations and their suitability for plastics which will come in contact with food make this type of stabilizer particularly interesting [8, 9].

The following organic tin compounds are particularly effective: Dialkyltin salts of fatty acids (for instance, lauric or stearic acid) or alpha, beta-unsaturated dicarboxylic acids, dialkyltin alkoxides and dialkyltin mercaptides. The tin compounds have a twofold function in stabilizing PVC: They are effective HCl acceptors which, due to complex formation, may even bind a larger than stoichometric amount of HCl:

Dibutyl-tin-dilaurate

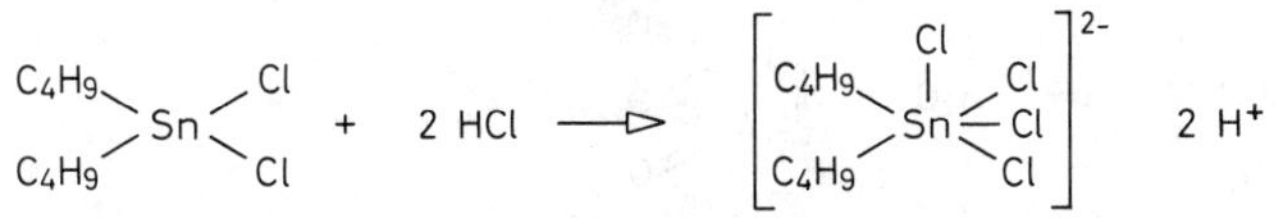

$$C_4H_9 \diagdown_{Sn} \diagup^{Cl}_{Cl} \diagup C_4H_9 \quad + \quad 2\ HCl \quad \longrightarrow \quad \left[C_4H_9 \diagdown_{Sn} \diagup^{Cl}_{Cl} \diagdown^{Cl}_{C_4H_9} \right]^{2-} \quad 2\ H^+$$

The second function of the organic tin stabilizer is the exchange of reactive chlorine atoms, which are responsible for the polymer degradation, for functional groups of the tin compound. It is primarily an exchange of an ester group for a chloride ion:

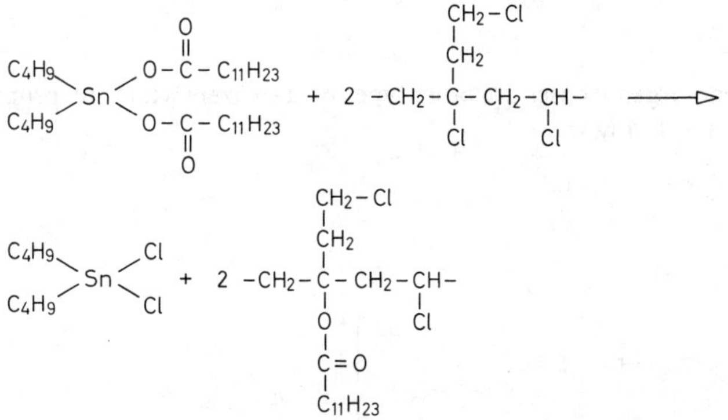

$$C_4H_9 \diagdown_{Sn} \diagup^{O-\overset{O}{\overset{\|}{C}}-C_{11}H_{23}}_{O-\underset{\|}{\underset{O}{C}}-C_{11}H_{23}} \quad + \quad 2 \ -CH_2-\overset{\overset{\displaystyle CH_2-Cl}{|}}{\underset{|}{\overset{|}{C}}}-CH_2-CH- \quad \longrightarrow$$
$$\qquad\qquad\qquad\qquad\qquad\qquad \underset{Cl}{} \qquad \underset{Cl}{}$$

$$C_4H_9 \diagdown_{Sn} \diagup^{Cl}_{Cl} \quad + \quad 2 \ -CH_2-\overset{\overset{\displaystyle CH_2-Cl}{|}}{\underset{\displaystyle CH_2}{}}-CH_2-CH-$$

Unstable sites are converted into stable chain structures in this manner, thereby interfering with the zipperlike HCl elimination. The organic tin compounds can also react with free radicals. Radical intermediates can be trapped in this manner:

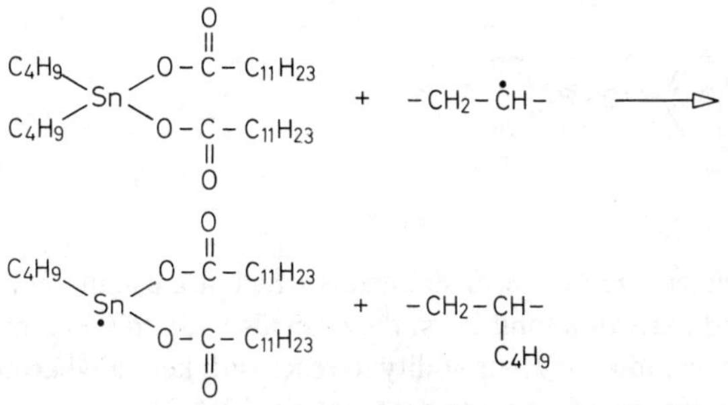

$$C_4H_9 \diagdown_{Sn} \diagup^{O-\overset{O}{\overset{\|}{C}}-C_{11}H_{23}}_{O-C-C_{11}H_{23}} \quad + \quad -CH_2-\overset{\bullet}{C}H- \quad \longrightarrow$$

$$C_4H_9 \diagdown_{\underset{\bullet}{Sn}} \diagup^{O-\overset{O}{\overset{\|}{C}}-C_{11}H_{23}}_{O-C-C_{11}H_{23}} \quad + \quad -CH_2-CH-$$

In this way, either radical reactions are blocked at the first stage or a reduction of the kinetic chain length of the radical chain reaction is achieved [1 a].

An important factor in the stabilization of PVC is the blocking of double bonds, in particular conjugated double bonds. Derivatives of maleic acid, which are used as stabilizers in the form of their metal salts, can change conjugated double bonds by a Diels-Alder reaction according to the following mechanism [1 a]:

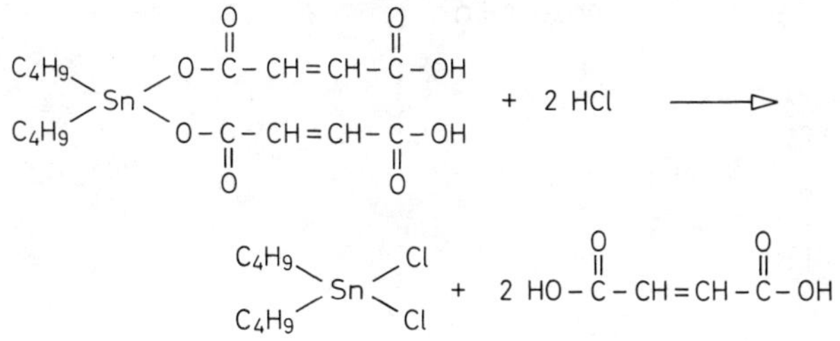

$$C_4H_9 \diagdown_{Sn} \diagup^{O-\overset{O}{\overset{\|}{C}}-CH=CH-\overset{O}{\overset{\|}{C}}-OH}_{O-\underset{\|}{\underset{O}{C}}-CH=CH-\underset{\|}{\underset{O}{C}}-OH} \quad + \quad 2\ HCl \quad \longrightarrow$$

$$C_4H_9 \diagdown_{Sn} \diagup^{Cl}_{Cl} \quad + \quad 2\ HO-\overset{O}{\overset{\|}{C}}-CH=CH-\overset{O}{\overset{\|}{C}}-OH$$

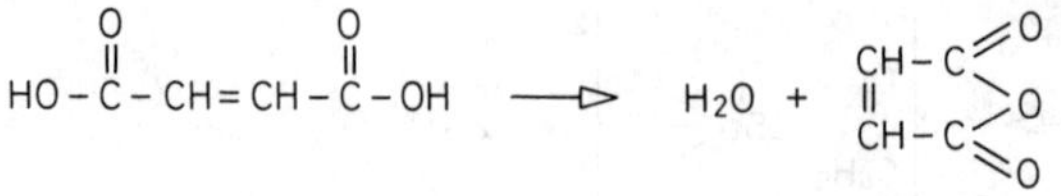

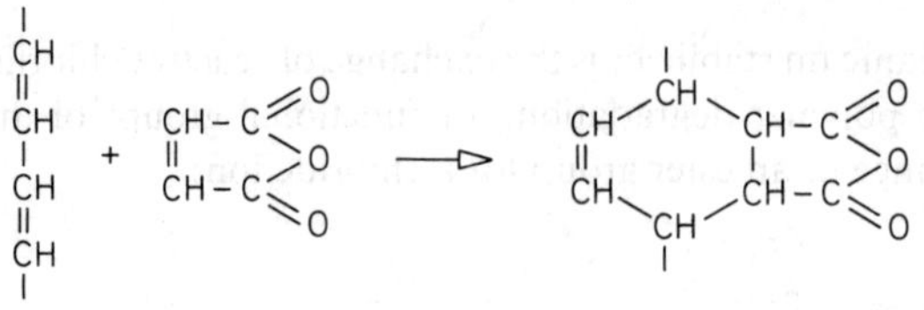

$$HO-\overset{\overset{O}{\|}}{C}-CH=CH-\overset{\overset{O}{\|}}{C}-OH \longrightarrow H_2O + \begin{matrix} CH-C \diagdown^{O}_{O} \\ \| \hspace{1cm} \diagup \\ CH-C \diagup_{O} \end{matrix}$$

$$\begin{matrix} | \\ CH \\ \| \\ CH \\ | \\ CH \\ \| \\ CH \\ | \end{matrix} + \begin{matrix} CH-C\diagup^{O} \\ \| \hspace{0.5cm} \diagdown_{O} \\ CH-C\diagdown_{O} \end{matrix} \longrightarrow \begin{matrix} \hspace{0.5cm} CH \\ CH \diagup \diagdown CH-C\diagup^{O} \\ \| \hspace{1.2cm} | \hspace{0.5cm} \diagdown \\ CH \diagdown \diagup CH-C\diagdown_{O} \\ \hspace{0.5cm} CH \end{matrix}$$

Epoxy compounds and organic bases are other types of stabilizers which act primarily as HCl acceptors and react as follows:

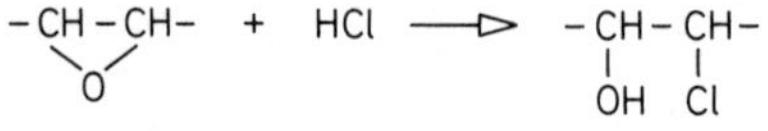

$$-CH-CH- + HCl \longrightarrow -CH-CH- \\ \diagdown_{O}\diagup \hspace{2cm} \underset{OH}{|} \hspace{0.3cm} \underset{Cl}{|}$$

Epoxy compound

$$S=C\diagup^{NH-C_6H_5}_{\diagdown NH-C_6H_5} + 2\,HCl \longrightarrow \left[S=C\diagup^{NH_2-C_6H_5}_{\diagdown NH_2-C_6H_5} \right]^{++} (Cl^-)_2$$

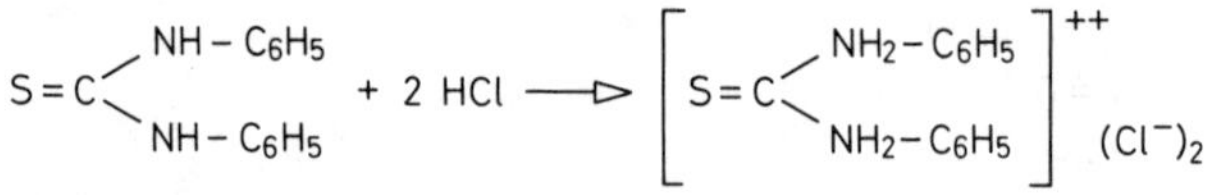

Diphenyl thiourea

Another type of stabilizer is phosphorus compounds, such as the bis(diphenyl phosphite) of bisphenol-A:

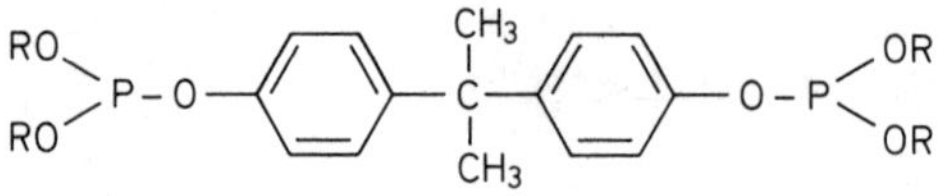

$$\begin{matrix} RO \\ \hspace{0.3cm}\diagdown \\ \hspace{0.6cm}P-O- \\ \diagup \\ RO \end{matrix} \langle\hspace{-0.2cm}\bigcirc\hspace{-0.2cm}\rangle \begin{matrix} CH_3 \\ | \\ -C- \\ | \\ CH_3 \end{matrix} \langle\hspace{-0.2cm}\bigcirc\hspace{-0.2cm}\rangle -O-P\diagup^{OR}_{\diagdown OR}$$

R = Phenyl

Phorphorus containing stabilizers are universal stabilizers which are usually not used alone. They are HCl absorbers and oxidation inhibitors; they can also add to C=C groups in degraded PVC and react with free radicals. Their ability to react with keto allyl groups is important; this will remedy the symptoms of polymer degradation [10, 11].

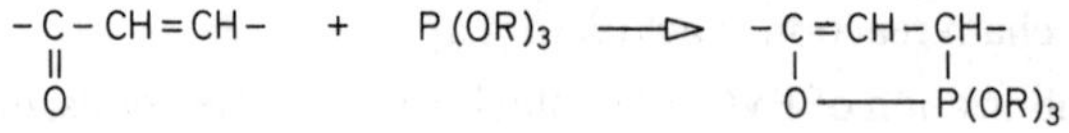

$$-\overset{\overset{|}{}}{\underset{\overset{\|}{O}}{C}}-CH=CH- + P(OR)_3 \longrightarrow -C=CH-CH- \\ \hspace{4.5cm} \underset{O-\!\!\!-\!\!\!-P(OR)_3}{|} \hspace{2cm} \text{(a)}$$

$$-\overset{\overset{O}{\|}}{C}-CH=CH- + P(OR)_3 \xrightarrow{HCl} -\overset{\overset{O}{\|}}{C}-CH_2-\underset{\underset{(RO)_2-P=O}{|}}{CH}- + RCl \hspace{1cm} \text{(b)}$$

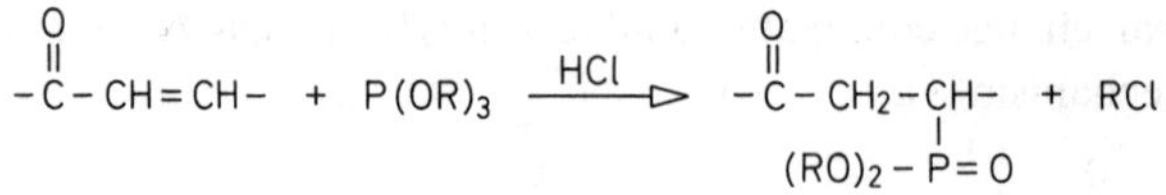

12.3 Degradation of Fluoro Polymers

Fluoro polymers, which do not contain hydrogen, behave on thermal exposure fundamentally different than PVC. The more halogen is replaced by hydrogen, the more similar to PVC these polymers become.

$$\left(\begin{array}{c} \overset{|}{\underset{|}{F}} \quad \overset{|}{\underset{|}{F}} \\ -C-C- \\ \overset{|}{F} \quad \overset{|}{F} \end{array}\right)_n$$

Polytetrafluoroethylene (PTFE) cannot be fabricated by the customary extrusion or injection molding processes, since, at the temperature necessary for processing in screw extruders, increased polymer degradation occurs. For this reason, the material is fabricated by ram extrusion or by sintering of preforms above the crystalline melting point.

PTFE decomposes at high temperatures above 400 °C and forms low molecular weight products. The composition of these products can be controlled by varying pressure, temperature and additives. Since the bond energy of the C-C bonds of the backbone is much less than that of the C-F bonds, cleavage at high temperature occurs almost exclusively at the backbone linkages. Only traces of fluorine can be found in the pyrolysate.

Contrary to widespread opinion, PTFE is combustible because ethylenic unsaturated pyrolysis gases can react with oxygen. At about 600 °C, PTFE burns, releasing a heat of combustion of about 4600 cal/g [12].

Oxidative decomposition of PTFE does not occur at processing temperatures and does not have to be taken into consideration by the processor.

Cross-linking of PTFE, either through a thermal mechanism or by radiation, is not possible, because the primary step of the energy impact is not the generation of chain radicals but of ethylenic unsaturated fragments. If PTFE is irradiated in the presence of oxygen, polar structures are formed at the surface as a result of polymer degradation and reaction with oxygen. This permits printing on the objects and makes them cementable.

Other perfluorinated polymers are similar to PTFE with respect to thermal decomposition and oxidative degradation. Only when fluorine is replaced by chlorine will the chains become more reactive.

$$\left(\begin{array}{c} \overset{|}{\underset{|}{F}} \quad \overset{|}{\underset{|}{F}} \\ -C-C- \\ \overset{|}{F} \quad \overset{|}{Cl} \end{array}\right)_n$$

Polytrifluorochloroethylene can be extruded and injection molded at temperatures of about 300 °C, but it does degrade rapidly at these temperatures. Like PTFE, it decomposes into low molecular weight compounds without any significant cross-linking. In contrast to PTFE, some elemental chlorine is released, which is extremely corrosive and places special requirements on the equipment and tools.

$$\left(\begin{array}{c} \overset{|}{\underset{|}{H}} \quad \overset{|}{\underset{|}{F}} \\ -C-C- \\ \overset{|}{H} \quad \overset{|}{F} \end{array}\right)_n$$

At high temperatures, *polyvinylidene fluoride* behaves similarly to PVC. Injection molding requires cylinder temperatures of about 275 °C. The pure polymer can be heated for 30 minutes to 340 °C without significant decomposition [13]. This allows a sufficient range for processing.

Like PVC, excessive temperatures cause elimination of hydrogen halide (in this case HF), which introduces a double bond into the backbone. This will activate adjacent fluorine atoms, which facilitates the elimination of additional hydrogen halide as it does in the case of PVC. Conjugated double bonds are formed by a zipperlike mechanism. The polymer discolors when the conjugated system has reached a sufficient length. Further reactions of those chain segments which are free of halogen result in chain cleavage and cross-linking. Hydrogen fluoride generated by the degradation of polyvinylidene fluoride is extremely corrosive.

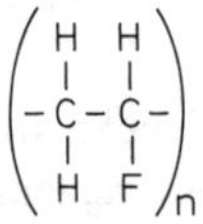

$$\left(\begin{matrix} H & H \\ | & | \\ -C - C- \\ | & | \\ H & F \end{matrix}\right)_n$$

Polyvinyl fluoride, just like PVC and polyvinylidene fluoride, tends to eliminate hydrogen halide.

12.4   Degradation of Polymethyl Methacrylate during Processing

If pure polymethyl methacrylate (PMMA) is heated to temperatures above about 180 °C, low molecular weight components are eliminated by a zipperlike mechanism. The rate of degradation is temperature-dependent.

During the polymerization of MMA, chain termination occurs, by combination as well as by disproportionation, with termination by disproportionation predominating. While combination results in a saturated compound, termination by disproportionation yields a molecule containing one saturated and one unsaturated end group:

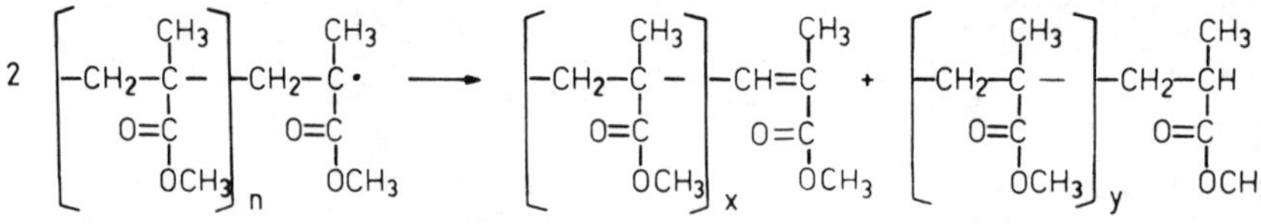

The terminal unsaturated component is unstable and, starting at one end, will split into the monomer components at temperatures higher than 180° [14]. The reaction can be monitored by measuring the weight loss, since the products generated by decomposition and the corresponding temperature are volatile (see Figure 41). Almost half of all polymer molecules have a terminal double bond because of chain termination being favored by disproportionation.

The saturated component is thermally more stable; a higher activation energy is needed for cleavage [16]. Only after chain cleavage has taken place does a zipperlike depolymerization occur.

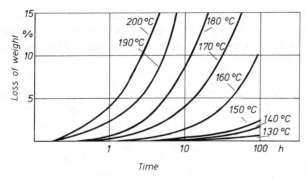

Figure 41   Loss of weight of PMMA Plexiglas® 240, sample size $30 \times 15 \times 3$ mm as a function of tempering time and temperature [15].

Blocking of terminal chain double bonds and incorporation of comonomer by means of copolymerization are methods to stabilize PMMA. If, for example, lauryl mercaptan is added onto the terminal chain double bonds, then the reactive (terminal chains) double bonds will be converted to the more stable thioether groups [17]:

$$\begin{array}{ccccc} CH_3 & CH_3 & & CH_3 & CH_3 \\ | & | & & | & | \\ -C-CH=C & + & HS-R & \longrightarrow & -C-CH_2-C-S-R \\ | & | & & | & | \\ O=C & O=C & & O=C & O=C \\ | & | & & | & | \\ OCH_3 & OCH_3 & & OCH_3 & OCH_3 \end{array}$$

Copolymerization of MMA with acrylic acid ester results in random incorporation of the ester into the polymer chain. Later, during processing, thermal decomposition will depolymerize the modified PMMA only up to the point where the free radical meets the acrylic acid ester unit [18]. The corresponding radical is more stable, and depolymerization stops. Further decomposition is possible only after chain cleavage and renewed formation of terminal radical methyl methacrylate groups.

Processing by injection molding of PMMA requires bulk temperatures of 150–230 °C depending on the type of polymer. Therefore, at processing temperatures polymer degradation and formation of volatile components are possible. The polymer processor has to be aware of these drawbacks. The polymer should stay in the hot part of the equipment for the shortest time possible.

12.5   Decomposition of Polystyrene during Processing

Like PMMA, polystyrene (PS) tends to regenerate monomer components during thermal degradation. But, only at temperatures above 200 °C does this decomposition occur and at such a rate that it is of practical importance.

Degradation of PS consists of two different partial reactions. During the first reaction the macromolecule splits into two free polymer radicals. It is believed that this does not involve a random decomposition but that degradation starts at selected "weak spots." These weak spots, for example, are segments with head-to-head addition, chain branch points, and unsaturated groups. Free chain radicals which are generated by chain cleavage at weak spots tend to split off into components of low molecular weight: monostyrene, and di-, tri-, and tetramers, with monomers being the dominant component. The rate of reaction of chain cleavage is of the zero order, while splitting off into components of low molecular weight seems to be of the first order [19].

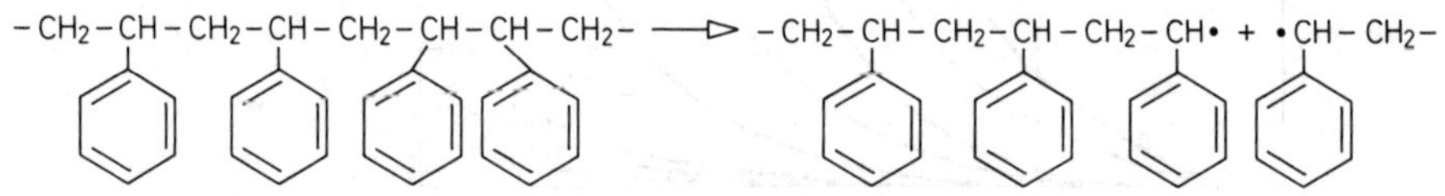

$$- CH_2 - CH - CH_2 - CH - CH_2 - CH - CH - CH_2 - \longrightarrow - CH_2 - CH - CH_2 - CH - CH_2 - CH\bullet + \bullet CH - CH_2 -$$

PS chain segment with
head-to-head structure

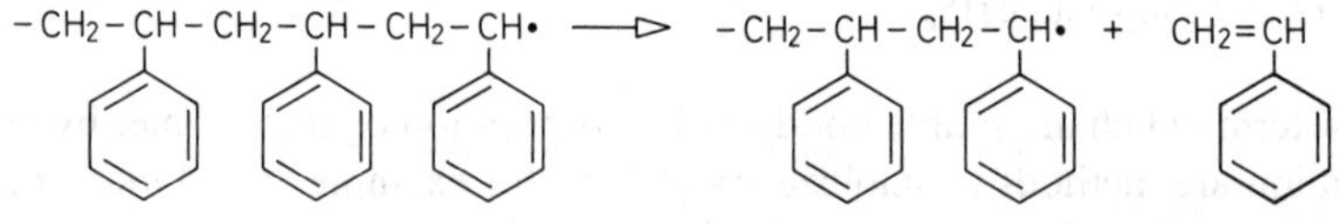

$$- CH_2 - CH - CH_2 - CH - CH_2 - CH\bullet \longrightarrow - CH_2 - CH - CH_2 - CH\bullet + CH_2 = CH$$

Depolymerization

Polymer degradation progresses in a zipperlike fashion, initiated by chain radicals which were generated through chain cleavage. Radicals which are alpha to the benzene ring are stabilized by mesomerism and their radical character may be partially transferred to the aromatic ring. This relative stability allows for greater kinetic chain lengths during depolymerization involving a radical chain reaction since radical transfer reactions are not significant.

Thermal decomposition of polystyrene is of little practical importance since the processing temperature is usually lower than the temperature at which polymer degradation is too fast to be tolerable. However, oxidative degradation of PS is possible during processing.

The reason for the relatively high oxidizability of PS is the resonance stabilization of radicals on the carbon atom which is alpha to the benzene ring. During processing with a screw extruder, oxidative chain degradation must be taken into account since oxygen gets into the melt accidentally. During oxidation the first rate-determining step is abstraction of a tertiary hydrogen atom (Eq. a). Oxygen is then added onto the newly generated free radical (Eq. b) followed by formation of a hydroperoxide (Eq. c), which will undergo random chain cleavage resulting in chain fragments containing a carboxyl group (Eq. d):

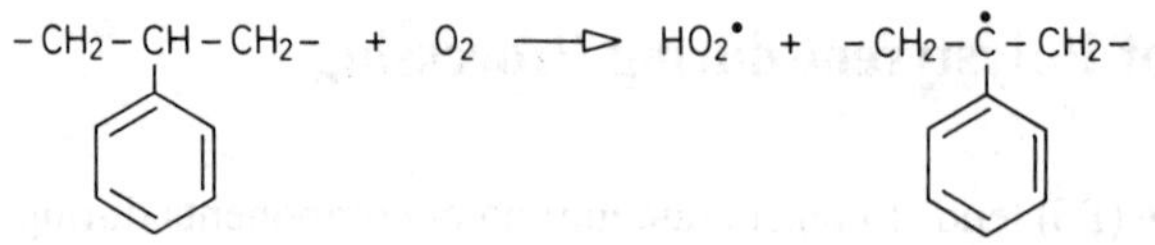

$$- CH_2 - CH - CH_2 - \quad + \quad O_2 \quad \longrightarrow \quad HO_2\bullet \quad + \quad - CH_2 - \overset{\bullet}{C} - CH_2 - \qquad\qquad\text{(a)}$$

$$- CH_2 - \overset{\bullet}{C} - CH_2 - \quad + \quad O_2 \quad \longrightarrow \quad - CH_2 - \overset{\overset{O\bullet}{|}\overset{|}{O}}{C} - CH_2 - \qquad\qquad\text{(b)}$$

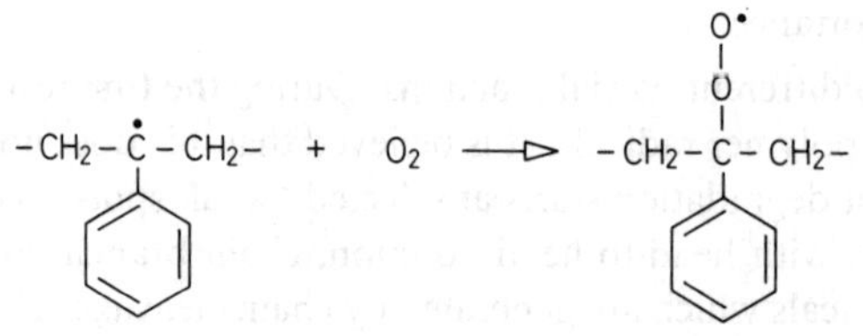

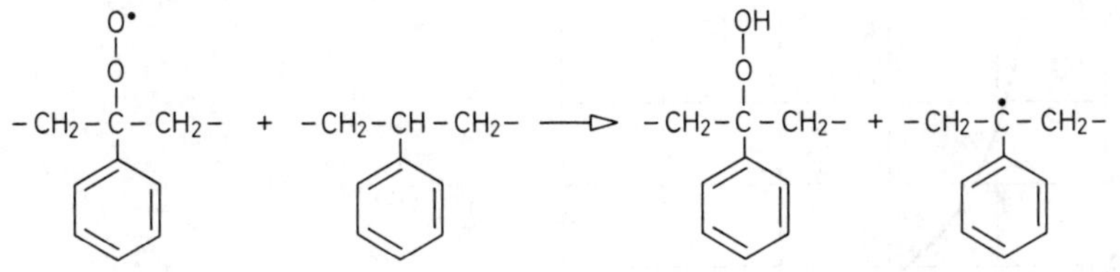

$$-CH_2-\overset{\overset{\displaystyle O^\bullet}{\overset{\displaystyle |}{O}}}{\underset{\underset{\bigcirc}{|}}{C}}-CH_2- \quad + \quad -CH_2-\underset{\bigcirc}{CH}-CH_2- \quad \longrightarrow \quad -CH_2-\overset{\overset{\displaystyle OH}{\overset{\displaystyle |}{O}}}{\underset{\underset{\bigcirc}{|}}{C}}-CH_2- \quad + \quad -CH_2-\overset{\displaystyle \bullet}{\underset{\bigcirc}{C}}-CH_2-$$

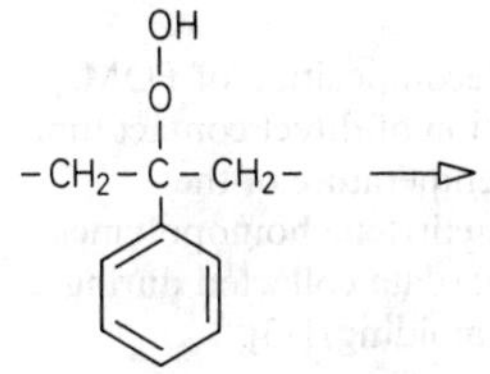

(c)

$$-CH_2-\overset{\overset{\displaystyle OH}{\overset{\displaystyle |}{O}}}{\underset{\underset{\bigcirc}{|}}{C}}-CH_2- \quad \longrightarrow \qquad \text{Thermal decomposition resulting in chain cleavage}$$

(d)

12.6   Degradation of Polyoxymethylene during Processing

Injection molding of polyoxymethylene (POM) takes place at bulk temperatures of 180–230 °C and mold temperatures of 60–120 °C. Since POM may already begin to decompose at these bulk temperatures (see Figure 42), the heating period of the molding compound has to be kept to a minimum to avoid extensive polymer degradation. Pure homopolymer is the most unstable polymer since, at elevated temperatures the depolymerization starts at the chain end.

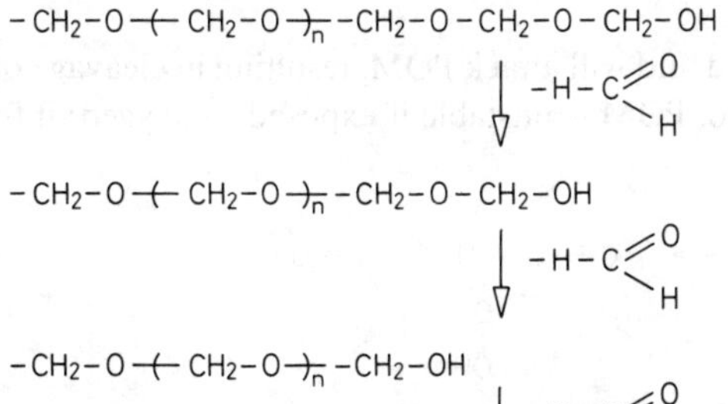

$$-CH_2-O-\!\!\left(\!CH_2-O\!\right)_{\overline{n}}-CH_2-O-CH_2-O-CH_2-OH$$
$$\downarrow \quad -H-C\!\!\underset{H}{\overset{O}{\diagup\!\!\!\backslash}}$$

$$-CH_2-O-\!\!\left(\!CH_2-O\!\right)_{\overline{n}}-CH_2-O-CH_2-OH$$
$$\downarrow \quad -H-C\!\!\underset{H}{\overset{O}{\diagup\!\!\!\backslash}}$$

$$-CH_2-O-\!\!\left(\!CH_2-O\!\right)_{\overline{n}}-CH_2-OH$$
$$\downarrow \quad -H-C\!\!\underset{H}{\overset{O}{\diagup\!\!\!\backslash}}$$

Stabilizers, copolymerization, and chemical alteration of the end groups will have a definite effect on the degradation of POM. For the time being, thermal depolymerization is inhibited by converting hemiacetal end groups into an ether or ester. Esterification with acetic acid anhydride is a commercially important method of blocking end groups. This reaction incorporates the acetic acid ester group into the polymer:

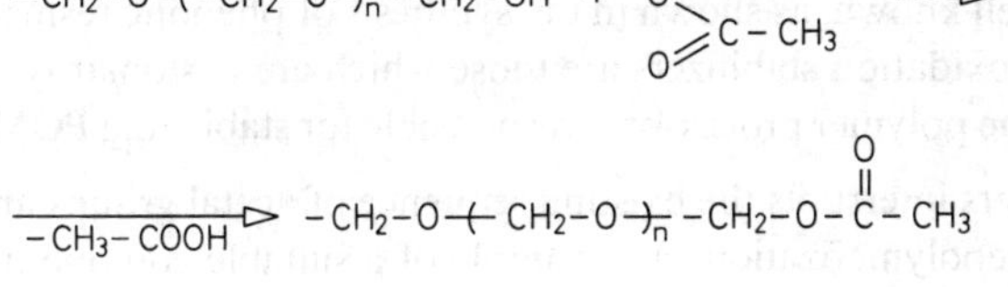

$$-CH_2-O-\!\!\left(\!CH_2-O\!\right)_{\overline{n}}-CH_2-OH \quad + \quad O\!\!\underset{\diagdown C-CH_3}{\overset{\diagup C-CH_3}{}} \quad \longrightarrow$$

$$\xrightarrow{-CH_3-COOH} \quad -CH_2-O-\!\!\left(\!CH_2-O\!\right)_{\overline{n}}-CH_2-O-\overset{\overset{\displaystyle O}{\|}}{C}-CH_3$$

Acetic acid ester end group

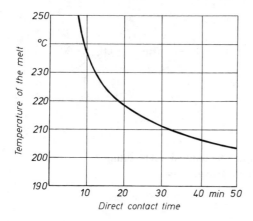

Figure 42
Start of decomposition of POM
as a function of direct contact time
and the temperature of the
polyoxymethylene homopolymer
melt (°C); (data collected during
injection molding) [13].

Hydrolysis of the ester groups restores the hydroxyl end groups, thus making thermal depolymerization possible. Cleavage of the ether group by acidic compounds will result in fragments containing hydroxyl end groups.

Formaldehyde formed during depolymerization is very sensitive to oxidation; oxygen will convert it to formic acid:

$$H-C\overset{\displaystyle O}{\underset{\displaystyle H}{\big<}} \quad + \quad 1/2\ O_2 \quad \longrightarrow \quad H-C\overset{\displaystyle O}{\underset{\displaystyle OH}{\big<}}$$

Formaldehyde   Formic acid

Formic acid is a relatively strong acid and will attack POM, resulting in cleavage of the ether groups and chain degradation. Also, POM is unstable if exposed to oxygen; it forms peroxide by adding onto oxygen:

$$-CH_2-O-CH_2-O-CH_2-O- \quad + \quad O_2 \quad \longrightarrow \quad -CH_2-O-\underset{\underset{\displaystyle OH}{\overset{\displaystyle |}{\underset{\displaystyle |}{O}}}}{CH}-O-CH_2-O-$$

At elevated temperatures the ether peroxides decompose and undergo chain cleavage. The degradation products will lead to accelerated chain depolymerization.

Initially, depolymerization can be prevented by blocking the end groups as discussed previously. Nevertheless, various reactions may cause formation of hydroxyl end groups which will allow depolymerization. The polymer can be made somewhat resistant to depolymerization by chemically linking the formaldehyde generated with other groups before it can be oxidized by oxygen. This prevents the formation of formic acid which is responsible for accelerating the degradation. Suitable stabilizers are phenols and urea; their addition reaction with formaldehyde is well known, as shown in the synthesis of phenolic resins and amino resins [13]. Also, suitable oxidation stabilizers are those which are customarily used for all other polymers. Usually, the polymer processor is responsible for stabilizing POM.

Incorporation of comonomers interrupts the existing sequence of acetal groups and is an effective way of preventing depolymerization. An example of a suitable comonomer is ethylene oxide, which will incorporate the following chain structure into the POM:

$$R-O-CH_2-O-CH_2-CH_2-(OCH_2)_n-O-CH_2-OH$$

Oxyethylene
unit

The thermal stability of POM with oxyethylene units is much higher than that of POM. Thermal depolymerization of a polyoxymethylene copolymer with unblocked chain ends will result in formation of a terminal oxyethylene unit. This will convert the end-position hemiacetal group into a thermally more stable alcoholic hydroxyl group; further degradation is possible only by preliminary chain cleavage caused by oxidation or acids which generates new acetal groups.

12.7 Degradation of Polyamides during Processing

Polyamides are synthesized either by polycondensation of a mixture of decarboxylic acids and diamine or by polymerization of a lactam.

The initial step is given by the following equilibrium:

$$R_1-COOH + H_2N-R_2 \rightleftharpoons R_1-\overset{\overset{\displaystyle O}{\|}}{C}-NH-R_2 + H_2O$$

Synthesis by polycondensation involves the law of mass action:

$$\frac{\left[R_1-\overset{\overset{\displaystyle O}{\|}}{C}-NH-R_2\right] \cdot \left[H_2O\right]}{\left[R_1-COOH\right] \cdot \left[H_2N-R_2\right]} = K$$

[ ] = concentration
K = equilibrium constant

Raw material and final product form an equilibrium during synthesis of polyamide. The equilibrium constant is dependent on the temperature.

Similar circumstances exist for the hydrolytic polymerization of ε-caprolactam; in the presence of minute amounts of water, ε-caprolactam will be converted into a compound of high molecular weight. The polymerization can be described by three equilibrium equations (see Section 6.7). During hydrolytic polymerization of ε-caprolactam, all components (water, ε-caprolactam, aminocaproic acid, oligomer, and polymer) are equilibrated. The equilibrium conditions are constantly maintained during storage of the raw material, during processing, and during application of the polyamides.

At low temperatures the equilibrium is reached very slowly; this makes polyamides semistable products at low temperatures even if they are not in a state of equilibrium. The processing of polyamides requires bulk temperatures of 230–290 °C depending on the type of polyamide. At these temperatures, the state of equilibrium is reached very quickly. A high water content in the polymer melt during processing will alter the polymer composition and result in noticeable variations in the properties.

At high temperatures, polyamide-6 which is low in monomer and oligomer, tends to regenerate compounds of low molecular weight by readjusting the equilibrium. Figure 43

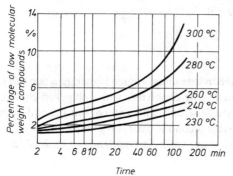

Figure 43
Regeneration of soluble compounds with low molecular weight in the polyamide 6 melt as a function of temperature and time [20].

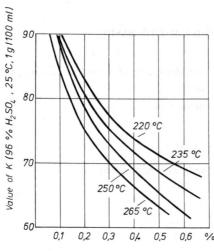

Figure 44
Equilibrium constant K of polyamide 6 melt as a function of temperature and water content [21].

shows the amounts of low molecular weight compounds regenerated at various temperatures and time intervals. Also, the viscosity of the solution of polyamide 6 changes rather quickly. Figure 44 shows the relative viscosity of the solution as a function of water content and temperature.

Since water is responsible for chain degradation during processing, the granular polyamide has to be dried by the polymer processor prior to processing, unless the material is delivered dry and in sealed containers.

Likewise, concentration and type of end groups are determining factors when the composition of polyamide is altered by means of water and heat.

During polymer synthesis the concentration of end groups can be controlled by additional monocarboxylic acids and monoamines. This modification incorporates into the polymer end groups which are no longer in equilibrium with other components.

$$-NH-\overset{\overset{\text{O}}{\|}}{C}-CH_3 \qquad\qquad\text{Blocked amino end group}$$

$$-\overset{\overset{\text{O}}{\|}}{C}-NH-CH_2-CH_2-CH_2-CH_3 \qquad\qquad\text{Blocked carboxylic end group}$$

A small number of end groups is equivalent to a low concentration of lactam during equilibrium. Therefore, a detrimental effect due to regeneration of monomer is less likely.

Polyamides are thermally very stable but are oxidized to a small degree at melting point temperatures. CH_2 groups which are next to an amino group represent the preferred point for initial attack. Therefore, at high temperatures, prolonged contact of polyamide with air is undesirable and will lead to discoloration (browning) of the surface.

12.8   Degradation of Polyethylene and Polypropylene during Processing

Polyethylene (PE) is thermally quite stable, chain degradation will take place only at high temperature. Since processing is done at comparatively low temperatures, strictly thermal degradation is very unlikely during processing. Polypropylene (PP) is also thermally quite stable at the usual processing conditions and does not undergo polymer degradation in the absence of air.

Strictly thermal degradation of PE will result in random chain cleavage (see Eq. a). The generated polymer radicals continue to react by producing fragments of low molecular weight (see Eqs. b, c) and by radical transfer (see Eq. d).

Initial attack at the chain:

$$- CH_2 - CH_2 - CH_2 - CH_2 - \longrightarrow \ - CH_2 - CH_2^\bullet \ + \ ^\bullet CH_2 - CH_2 - \tag{a}$$

Degradation reaction:

$$- CH_2 - CH_2 - CH_2 - CH_2^\bullet \longrightarrow \ - CH_2 - CH_2^\bullet \ + \ CH_2 = CH_2 \tag{b}$$

$$- \overset{\bullet}{C}H - CH_2 - CH_2 - CH_2 - \longrightarrow \ - CH = CH_2 \ + \ ^\bullet CH_2 - CH_2 - \tag{c}$$

Radical transfer:

$$- CH_2 - CH_2^\bullet \ + \ - CH_2 - \overset{\overset{\displaystyle H}{|}}{C}H - CH_2 - CH_2 - \longrightarrow \ - CH_2 - CH_3 \ + \ - CH_2 - \overset{\bullet}{C}H - CH_2 - CH_2 - \tag{d}$$

PE and PP are comparable in their thermal stability, but they react easily with oxygen and free radicals, resulting in alteration of polymer properties. A total exclusion of oxygen is impossible, since the polymer contains a certain amount of dissolved oxygen.

At higher temperatures oxygen will act as a biradical and generate free chain radicals which may react in the following manner (see Chapter 9):

$$- CH_2 - CH_2 - CH_2 - \ + \ O_2 \longrightarrow \ - CH_2 - \overset{\bullet}{C}H - CH_2 - \ + \ HO_2^\bullet$$

Starting reaction   Free (polymer)
                                    radical

$$- CH_2 - \overset{\bullet}{C}H - CH_2 - \ + \ O_2 \longrightarrow \ - CH_2 - \overset{\overset{\displaystyle \bullet O}{\overset{\displaystyle |}{\displaystyle O}}}{\underset{|}{C}}H - CH_2 -$$

Peroxide free radical

$$\begin{array}{c} O^\bullet \\ | \\ O \\ | \end{array} \qquad\qquad\qquad \begin{array}{c} OH \\ | \\ O \\ | \end{array}$$
$$-CH_2-CH-CH_2- + -CH_2-CH_2-CH_2- \longrightarrow -CH_2-CH-CH_2- + -CH_2-\overset{\bullet}{C}H-CH_2-$$

Hydroperoxide

Dimerization of various free radicals will stop the chain propagation:

$$\begin{array}{c} -CH_2-CH-CH_2 \\ \bullet \\ + \\ \bullet \\ -CH_2-\overset{\bullet}{C}H-CH_2 \end{array} \longrightarrow \begin{array}{c} -CH_2-CH-CH_2- \\ | \\ -CH_2-CH-CH_2- \end{array} \qquad \text{Cross-linking}$$

$$\begin{array}{c} -CH_2-CH-CH_2- \\ \bullet \\ + \\ O\bullet \\ | \\ O \\ | \\ -CH_2-C-CH_2- \end{array} \longrightarrow \begin{array}{c} -CH_2-CH-CH_2- \\ | \\ O \\ | \\ O \\ | \\ -CH_2-CH-CH_2- \end{array} \qquad \begin{array}{l}\text{Cross-linking by formation} \\ \text{of dialkyl peroxide}\end{array}$$

While cross-linking by formation of a C-C bond results in a relatively stable compound, dialkyl peroxides are unstable compounds which decompose to free radicals at processing temperatures and again intervene in the oxidation reaction:

$$\begin{array}{c} -CH_2-CH-CH_2- \\ | \\ O \\ | \\ O \\ | \\ -CH_2-CH-CH_2 \end{array} \longrightarrow \quad 2 \begin{array}{c} O^\bullet \\ | \\ -CH_2-CH-CH_2- \end{array}$$

$$\begin{array}{c} O^\bullet \\ | \end{array} \qquad\qquad\qquad \begin{array}{c} OH \\ | \end{array}$$
$$-CH_2-CH-CH_2- + -CH_2-CH_2-CH_2- \longrightarrow -CH_2-CH-CH_2- + -CH_2-\overset{\bullet}{C}H-CH_2-$$

Polyethylene   Free polymer radical

This newly formed hydroxyl group activates the methylene groups in the alpha position; therefore, the reaction product is less stable to oxidation than PE itself is.

At processing temperature, hydroperoxides decompose and form a free hydroxyl radical and a free polymer radical.

$$\begin{array}{c} -CH_2-CH-CH_2- \\ | \\ O \\ | \\ OH \end{array} \longrightarrow \begin{array}{c} -CH_2-CH-CH_2- + \, ^\bullet OH \\ | \\ O^\bullet \end{array}$$

$$-CH_2-CH_2-CH_2- + \, ^\bullet OH \longrightarrow -CH_2-\overset{\bullet}{C}H-CH_2- + H_2O$$

The free hydroxyl radical, as well as the free chain radical, start further chain reaction by attacking another PE chain, thereby losing their free radical character.

In addition to these free radical reactions, hydroperoxides are able to form ketones by eliminating water:

$$-CH_2-\overset{\overset{\displaystyle OH}{\overset{\displaystyle |}{\underset{\displaystyle |}{O}}}}{CH}-CH_2- \quad\longrightarrow\quad -CH_2-\overset{\overset{\displaystyle O}{\displaystyle \|}}{C}-CH_2- + H_2O$$

Ketone

The decomposition of free polymer radicals and the oxidation of a methylene group, in the alpha position to a keto group will result in chain degradation:

$$-CH_2-\overset{\overset{\displaystyle O}{\displaystyle \|}}{C}-CH_2- + O_2 \quad\longrightarrow\quad -CH_2-\overset{\overset{\displaystyle O}{\displaystyle \|}}{C}-OH + O=CH-$$

$$-CH_2-\overset{\overset{\displaystyle O^\bullet}{\displaystyle |}}{CH}-CH_2- \quad\longrightarrow\quad -CH_2-C\overset{\displaystyle \nearrow O}{\underset{\displaystyle \searrow H}{}} + {}^\bullet CH_2-$$

While oxidation without chain degradation will result only in formation of functional groups containing oxygen which will provide the polymer with slightly different properties, chain degradation will lead to deterioration of polymer quality.

In the presence of oxygen, PE and PP behave differently during processing. The bond energy of hydrogen atoms decreases from primary to tertiary hydrogen atoms. PP contains a high percentage of tertiary hydrogen atoms which easily provide sites for the initiation of the degradation process at high temperatures. Therefore, under comparable conditions, PP absorbs oxygen much faster than PE does.

Polypropylene hydroperoxides produced by oxidation will decompose easily, resulting in chain degradation.

$$-CH_2-\overset{\overset{\displaystyle OH}{\overset{\displaystyle |}{\overset{\displaystyle O}{\overset{\displaystyle |}{\underset{\underset{\displaystyle CH_3}{\displaystyle |}}{C}}}}}}{}-CH_2-\overset{}{\underset{\underset{\displaystyle CH_3}{\displaystyle |}}{CH}}- \quad\longrightarrow\quad -CH_2-\overset{\overset{\displaystyle O^\bullet}{\displaystyle |}}{\underset{\underset{\displaystyle CH_3}{\displaystyle |}}{C}}-CH_2-\overset{}{\underset{\underset{\displaystyle CH_3}{\displaystyle |}}{CH}}- + {}^\bullet OH$$

$$-CH_2-\overset{\overset{\displaystyle O^\bullet}{\displaystyle |}}{\underset{\underset{\displaystyle CH_3}{\displaystyle |}}{C}}-CH_2-\overset{}{\underset{\underset{\displaystyle CH_3}{\displaystyle |}}{CH}}- \quad\longrightarrow\quad -CH_2-\overset{\overset{\displaystyle O}{\displaystyle \|}}{C}-CH_3 + {}^\bullet CH_2-\overset{}{\underset{\underset{\displaystyle CH_3}{\displaystyle |}}{CH}}-$$

Ketone

$${}^\bullet CH_2-\overset{}{\underset{\underset{\displaystyle CH_3}{\displaystyle |}}{CH}}- \quad\longrightarrow\quad CH_3-\overset{\overset{\displaystyle \bullet}{}}{\underset{\underset{\displaystyle CH_3}{\displaystyle |}}{C}}-$$

Tertiary free radical

Cleavage of the hydroperoxide will result in formation of an oxonium radical which converts by chain degradation to a stable ketone and a relatively stable tertiary free radical.

Polypropylene is much more predisposed to chain degradation than polyethylene; this has been discussed earlier in chapters about cross-linking. While a high percentage of free PE radicals are stabilized by dimerization (cross-linking) the corresponding free polypropylene radical decomposes easily, resulting in chain degradation:

$$-CH_2-\overset{\bullet}{\underset{\underset{CH_3}{|}}{C}}-CH_2-\underset{\underset{CH_3}{|}}{CH}- \longrightarrow -CH_2-\underset{\underset{CH_2}{\|}}{CH} + \overset{\bullet}{CH_2}-\underset{\underset{CH_3}{|}}{CH}- \longrightarrow CH_3-\overset{\bullet}{\underset{\underset{CH_3}{|}}{C}}-$$

It is possible to vary the degree of degradation of PE and PP by using reducing agents and a free radical scavenger. The polymer manufacturer is responsible for incorporating these components into the molding compound.

Bibliography to Chapter 12

[1] a) *Voigt, J.:* »Die Stabilisierung der Kunststoffe gegen Licht und Wärme«, Springer Verlag, Berlin-Heidelberg-New York, 1966.
   b) *Gould, R. F.:* »Advances in Chemistry«, Serie 85, »Stabilization of Polymers and Stabilizer Processes«, Amer. chem. Soc., Washington D.C., 1968.

[2] *Braun, D.:* Pure Appl. Chem. 26 (1971) 173.

[3] *Hemmer, E., H. Jesse, H.-Ch. Rhiem:* Gummi, Asbest, Kunstst. 29 (1976) 204.

[4] *Braun, D., W. Quarg:* Angew. Makromol. Chem. 29/30 (1973) 163.

[5] *Owen, E. D., R. L. Read:* J. Polym. Sci., Polym. Chem. Ed. 17 (1979) 2719.

[6] *Minsker, K. S.:* Plaste Kautsch. 24 (1977) 375.

[7] *Reichherzer, R.:* Kunststoffe-Plastics 6 (1959) 90.

[8] *Ghatge, N. D., S. V. Vaidya:* Kautsch., Gummi, Kunstst. 32 (1979) 254.

[9] *Klimsch, P.:* Plaste Kautsch. 24 (1977) 380.

[10] *Kirpitschnikow, P. A., D. G. Pobedimski:* Plaste Kautsch. 22 (1975) 400.

[11] *Svoboda, P., F. Erben, R. Vesely:* Plaste Kautsch. 23 (1976) 23.

[12] *Duns, H. C.:* Ind. Engng. Chem. 47 (1955) 1445.

[13] *Vieweg, R., M. Reiher, H. Scheurlen:* »Polyacetale, Epoxidharze, fluorhaltige Polymerisate, Silicone usw.«, Kunststoff-Handbuch, Bd. XI, Carl Hanser Verlag, München, 1971.

[14] a) *Grassie, N.:* »Chemistry of High Polymer Degradation Processes«, Butterworth Scientific Publications, London.
   b) *Grassie, N., H. W. Mellville:* Proc. Roy. Soc. A., 199 (1949) 1.

[15] *Vieweg, R., F. Esser:* »Polymethylmethacrylate«, Kunststoff-Handbuch, Bd. IX, Carl Hanser Verlag, München, 1975.

[16] *Loshaek, S.:* J. Polym. Sci. 15 (1955) 391.

[17] US-PS 2.462.895 (1945), Du Pont.

[18] DE-RP 706.177 (1939), Röhm GmbH.

[19] *Madorsky, S. L.:* »Thermal Degradation of Organic Polymers«, Interscience Publishers, New York, 1964.

[20] *Vieweg, R., A. Müller:* »Polyamide«, Kunststoff-Handbuch, Bd. VI, Carl Hanser Verlag, München.

[21] *Reimschüssel, H. K.:* J. Polym. Sci. 41 (1959) 457.

Index